Springer Tracts in Advanced Robotics 89

Editors

Prof. Bruno Siciliano
Dipartimento di Ingegneria Elettrica
e Tecnologie dell'Informazione
Università degli Studi di Napoli
Federico II
Via Claudio 21, 80125 Napoli
Italy
E-mail: siciliano@unina.it

Prof. Oussama Khatib
Artificial Intelligence Laboratory
Department of Computer Science
Stanford University
Stanford, CA 94305-9010
USA
E-mail: khatib@cs.stanford.edu

For further volumes:
http://www.springer.com/series/5208

STAR (Springer Tracts in Advanced Robotics) has been promoted under the auspices of EURON (European Robotics Research Network)

Jürgen Sturm

Approaches to Probabilistic Model Learning for Mobile Manipulation Robots

Springer

Dr. Jürgen Sturm
Technische Universität München
Institute of Computer Science
Computer Vision Group
Boltzmannstraße 3
85748 Garching
Germany

ISSN 1610-7438 ISSN 1610-742X (electronic)
ISBN 978-3-642-37159-2 ISBN 978-3-642-37160-8 (eBook)
DOI 10.1007/978-3-642-37160-8
Springer Heidelberg New York Dordrecht London

Library of Congress Control Number: 2013936730

Printed on acid-free paper

Springer is part of Springer Science+Business Media (www.springer.com)

Für Julia und meine Familie

Editorial Foreword

Robotics is undergoing a major transformation in scope and dimension. From a largely dominant industrial focus, robotics is rapidly expanding into human environments and vigorously engaged in its new challenges. Interacting with, assisting, serving, and exploring with humans, the emerging robots will increasingly touch people and their lives.

Beyond its impact on physical robots, the body of knowledge robotics has produced is revealing a much wider range of applications reaching across diverse research areas and scientific disciplines, such as: biomechanics, haptics, neurosciences, virtual simulation, animation, surgery, and sensor networks among others. In return, the challenges of the new emerging areas are proving an abundant source of stimulation and insights for the field of robotics. It is indeed at the intersection of disciplines that the most striking advances happen.

The *Springer Tracts in Advanced Robotics (STAR)* is devoted to bringing to the research community the latest advances in the robotics field on the basis of their significance and quality. Through a wide and timely dissemination of critical research developments in robotics, our objective with this series is to promote more exchanges and collaborations among the researchers in the community and contribute to further advancements in this rapidly growing field.

The monograph by Jürgen Sturm is a contribution aimed at developing novel learning techniques enabling mobile manipulation robots to autonomously adapt to unstructured environments. Original solutions are proposed to the problem of inferring the kinematic structure of the robot as well as that of the articulated objects the robot interacts with from its own perception, including both visual and tactile data. The resulting integrated system allows teaching-by-demonstration of manipulation tasks by human operators. Probabilistic modelling techniques and efficient machine learning algorithms are thoroughly tested not only in simulation but also in a reach set of experiments in realistic operating conditions throughout the chapters of this volume.

Remarkably, the monograph is based on the author's doctoral thesis, which won the 2011 ECCAI Artificial Intelligence Dissertation Award and was a runner-up for the 2012 EURON Georges Giralt PhD Award. A very fine addition to STAR!

Naples, Italy
February 2013

Bruno Siciliano
STAR Editor

Foreword

The ability to robustly learn models is one of the fundamental pre-conditions for autonomous service robots. The knowledge about how objects in the environment can be used, which activity the user is carrying out, and how the robot functions are typical examples of relevant models that greatly facilitate the execution of corresponding tasks, in particular in the context of mobile manipulation.

In this book by Jürgen Sturm, the reader will find innovative and pioneering solutions to the problem of inferring the kinematic structure of the robot or of objects from its own perceptions. It furthermore describes how objects can be better detected by interacting with them. Finally, it presents an approach to learn object manipulation tasks based on imitation.

I know Jürgen since more than six years now and I always appreciated his enthusiasm. Jürgen has been working on a large variety of tasks and robots and collaborated with numerous students and colleagues. He is able to fascinate people to work on the topics he is interested in and this variety and fascination can be directly felt when reading this book. Jürgen has published his work in several international conferences and journals. These publications, and this is worth to be mentioned, have appeared in three different fields including robotics, artificial intelligence and neuroscience. He has received the ECCAI Dissertation award 2012 and was runner-up for the EURON Dissertation Award in 2012.

This book is an extensive introduction to state-of-the-art technology in robotic self-perception and inference of kinematic structures. The reader will find a rigorous probabilistic approach coupled with modern machine learning techniques. The book focuses on the probabilistic modeling and efficient algorithms for inference and demonstrates the resulting capabilities in real robotic experiments. A further strength lies in the sound and thorough evaluation of all presented techniques going beyond the world of simulation. The book is furthermore nicely written and contains plenty of illustrative figures. This makes this book readable not only for researchers but also for students who want to understand more about this exciting field. And finally, the book can furthermore serve as a source for wonderful classroom material.

Freiburg, Germany
January 2013

Wolfram Burgard

Foreword

Preface

Mobile manipulation robots are envisioned to provide many useful services both in domestic environments as well as in the industrial context. Examples include domestic service robots, that implement large parts of the housework, and versatile industrial assistants, that provide automation, transportation, inspection, and monitoring services. The challenge in these applications is that the robots have to function under changing, real-world conditions, be able to deal with considerable amounts of noise and uncertainty, and operate without the supervision of an expert. To meet these challenges, current robotic systems are typically custom-tailored to specific applications in well-defined environments, and therefore cannot deal robustly with changes in the situation. This book presents novel learning techniques that enable mobile manipulation robots, i.e., mobile platforms with one or more robotic manipulators, to autonomously adapt to new or changing situations. The approaches presented in this book cover the following four topics: (1) learning the robot's kinematic structure and properties using actuation and visual feedback, (2) learning about articulated objects in the environment in which the robot is operating, (3) using tactile feedback to augment the visual perception, and (4) learning novel manipulation tasks from human demonstrations.

In the first part of this book, we present innovative approaches to learning a robot's own body schema from scratch using visual self-observation. This allows manipulation robots to calibrate themselves automatically and to adapt their body schemata autonomously, for example after hardware failures or during tool use. In the second part, we extend the developed framework to learning about articulated objects – such as doors and drawers – with which service robots often need to interact. The presented algorithms enable robots to learn accurate kinematic models of articulated objects, which in turn allow them to interact with the objects robustly. In the third part, we provide approaches that allow manipulation robots to make use of tactile perception – an ability that is known to play an important role in human object manipulation skills. The main contributions in this part are approaches to identifying objects and to perceiving aspects of their internal states. With this, a manipulation robot can verify that it has grasped the correct object and, for example, discriminate full from empty bottles. Finally, we present an integrated system that allows human operators to intuitively teach a robot novel manipulation tasks by demonstration.

All techniques presented in this book are based on probabilistic learning and inference. They have been implemented and evaluated on real robots as well as in simulation. Extensive experiments have been conducted to analyze and validate the properties of the developed algorithms and to demonstrate a significant increase in robustness, adaptability, and utility of mobile manipulation robots in everyday life.

Acknowledgements

I would like to thank all the wonderful people in our lab at the University of Freiburg. This work would never have been possible without the inspiration and continuous support of my advisor Wolfram Burgard, who provided me with the right balance of encouragement, practical guidance, and opportunities, and granted me an exceptional degree of freedom in pursuing my own ideas. I thank my co-advisor Kurt Konolige for strengthening my view on real-world perception problems in robotics and making my research stay at Willow Garage possible. After my advisors, my sincere thank goes to Cyrill Stachniss and Christian Plagemann for the excellent conversations we had over the years. Both supported me countless times with good advice and contributed valuable ideas to this work.

Next, I thank my co-authors for the insightful discussions, fruitful collaborations, and late-night paper writing sessions. In particular, I would like to thank Advait Jain, Charlie Kemp, and Vijay Pradeep for our joint works on articulated objects; Sachin Chitta, Matthew Piccoli, Alexander Schneider, Marco Reisert, and Hans Burkhardt for our initiatives on tactile perception; and Clemens Eppner and Maren Bennewitz for our work on imitation learning. It was a pleasure for me to work with all of them.

Furthermore, I would like to thank the people at Willow Garage and all contributors to the robot middleware ROS – I am convinced that these efforts have significantly brought the robotics community forward. Thanks also to Kevin O'Regan who elucidated me on the philosophical and psychological aspects of self-perception and body schema learning. I thank Armin Hornung, Andreas Karwath, Axel Rottmann, Barbara Frank, Christoph Sprunk, Daniel Meyer-Delius, Daniela Sturm, Dominik Joho, Felix Endres, Henrik Kretzschmar, Julia Frankenberger, Jürgen Hess, Kai Wurm, Lionel Ott, Marit van Dijk, and Maximilian Beinhofer for proof-reading earlier versions of this document. Also, I would like to thank Susanne Bourjaillat, Kris Haberer, and Michael Keser for their administrative and technical support during my time in Freiburg.

My deepest gratitude goes to my family for the support and love they gave me in every period of my life. Finally, I thank Julia for her love during all these years.

This work has partly been supported by the German Research Foundation (DFG) through the Leibniz program, the intern program of Willow Garage, the European Commission under grant agreement numbers FP6-IST-004250-COSY and FP6-IST-045388-INDIGO, and by the German Ministry for Education and Research (BMBF) through the DESIRE project.

Acknowledgements

Contents

List of Figures

List of Tables

List of Algorithms

Chapter 1
Introduction

The development of flexible mobile manipulation robots is widely envisioned as a large breakthrough in technology and is expected to have a significant impact on our economy and society in the future. Mobile manipulation robots that are equipped with one or more gripper arms could fulfill various useful services in private homes such as cleaning, tidying up, or cooking, which would mean a significant time benefit to their owners. By supporting elderly and mobility-impaired people in the activities of daily living, such robots can reduce the dependency on external caregivers and support them to live a self-determined and autonomous life. Small and medium-sized enterprises would profit enormously from robotic co-workers that they can easily reconfigure to new production tasks. This technology would significantly lower the production costs of smaller companies and thus provide them with a significant competitive advantage. The goal of this book is to present novel approaches that enable mobile manipulation robots to be flexibly used in everyday life. The challenge in these applications is that robots operating in unstructured environments have to cope with less prior knowledge about themselves and their surroundings. Therefore, they need to be able to autonomously learn suitable probabilistic models from their own sensor data to robustly fulfill their tasks.

For decades, stationary manipulation robots have successfully been used in industrial mass production. In these applications strong assumptions about the physical setup and a controlled environment allow the creation of efficient but highly engineered approaches. These solutions are custom-tailored to specific applications which makes them difficult to adapt: typically, changes in the application require the manual adaptation of the robot's control code, a new layout of its work cell, and possibly the reconfiguration of its hardware. For this reason, industrial manipulators require the supervision of experts on a regular basis, and are therefore only cost-effective for the mass production. In contrast, the environment of mobile manipulators used for domestic service tasks or in small series production is largely unstructured, i.e., it can neither be exactly specified nor easily controlled. To deal with these uncertainties, mobile manipulation robots need to be considerably more flexible, robust, and adaptive than their stationary predecessors.

To illustrate the relevance of the topics presented in this book, we motivate our work using a typical example task of a domestic service robot. We assume that the robot is given the task to deliver a drink, which requires the robot

J. Sturm: *Approaches to Probabilistic Model Learning*, STAR 89, pp. 1–11.
DOI: 10.1007/978-3-642-37160-8_1

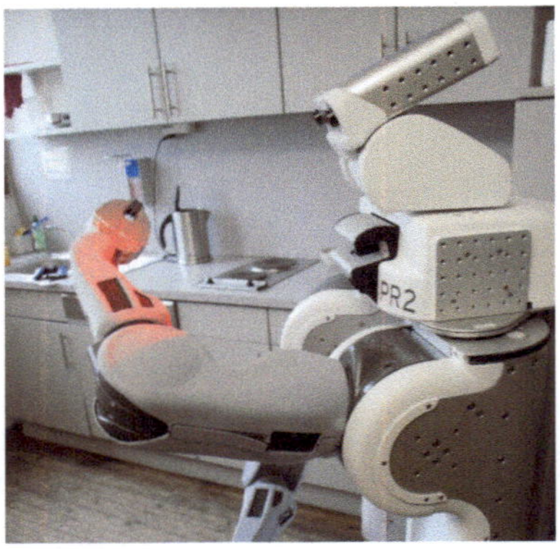

(a) body schema learning

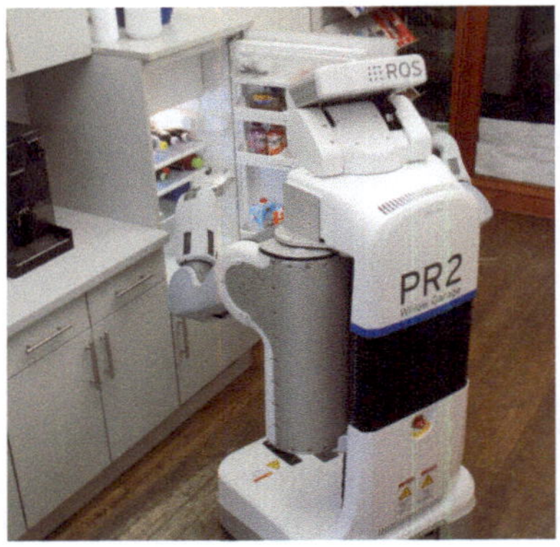

(b) articulated objects

(c) tactile sensing

(d) imitation learning

Fig. 1.1 Illustration of the four research questions addressed in this book. (a) Body schema learning using visual self-observation. (b) Learning to operate articulated objects, here: a fridge. (c) Using tactile sensing to estimate the state of a container. (d) Imitation learning to acquire novel manipulation skills.

to open the fridge, pick up the right bottle, and pour its content into a glass. To be able to accurately use its manipulator, the robot first needs to verify its body schema using visual self-observation (Figure 1.1a). This enables the robot to compensate for mechanical inaccuracies and detect potential hardware failures. Robots operating over extended periods of time without regular inspection would otherwise become increasingly inaccurate – and fail to accomplish their tasks. Once the robot established its body schema, it navigates to the fridge to retrieve a drink (Figure 1.1b). To open the fridge, the robot identifies the fridge door and generates a suitable trajectory for opening it. This, in turn, requires a kinematic model of the fridge. The robot learns this model from its observations and uses it subsequently to operate the door. Being able to learn kinematic models is fundamental for versatile service robots, as there are too many different cabinet doors and drawers in domestic environments to exclusively rely on predefined models. After the robot has successfully opened the fridge, it picks up a bottle. By using its tactile sensors, the robot can then detect whether the grasped bottle is full or empty (Figure 1.1c). This additional source of information about the object being manipulated is important because it enables a robot to verify that it has grasped the correct object and that this object is in the expected state. The next step of the delivery task is to pour the drink into a glass (Figure 1.1d). This skill, however, might not be part of the robot's current programming. In this case, the user can teach the robot this novel manipulation skill by demonstrating it to the robot. From this demonstration, the robot learns and generalizes a description of the task that it can subsequently use to reliably reproduce it – even for different positions of the glass and the bottle. Such an intuitive programming interface is an essential prerequisite for the usability of service robots in everyday life.

This motivating example leads us to the four research questions that we investigate in this book:

- How can a manipulation robot learn to accurately position its arm?
- How can a manipulation robot robustly operate doors and drawers?
- How can a manipulation robot infer the state of the objects it manipulates?
- How can a user intuitively teach novel manipulation tasks to a robot?

A robot that operates in unstructured environments with no or minimal human supervision needs to be able to perceive the world through its own sensors, and subsequently, build from this data an internal, up-to-date representation of the world. As sensor data is always noisy and potentially incomplete, a robot requires robust techniques to interpret and integrate it intelligently into its own models of the world. A robot can then use these models to estimate the state of objects in the world, simulate the consequences of its actions, generate plans, and, finally, verify the success of its actions.

In sum, this book provides novel probabilistic learning techniques that enable a manipulation robot

- to learn the body schema of its arm from scratch using self-observation, and to monitor and adapt this model over extended periods of time,
- to learn kinematic models of articulated objects from observation or interaction to reliably operate doors and drawers,
- to learn tactile object models to estimate the identity and state of the objects being manipulated, and
- to learn novel manipulation tasks from human demonstrations, and to reproduce them robustly in similar situations.

This book is organized as follows. In Chapter 2, we provide the technical background in machine learning and probabilistic modeling that we require in the remainder of this book. In particular, we introduce regression and

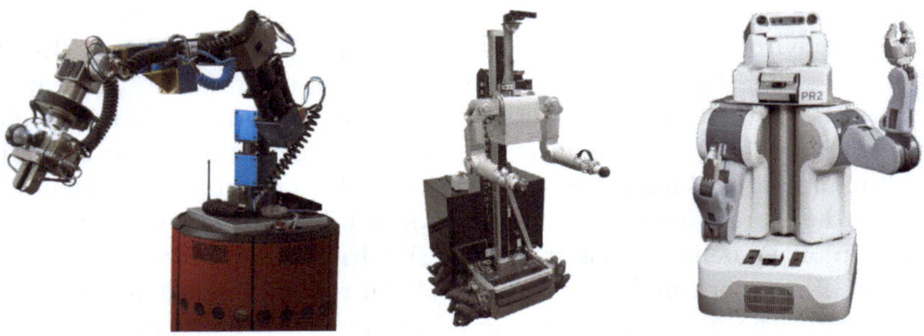

(a) Zora (b) Cody (c) Marvin

Fig. 1.2 Three state-of-the-art mobile manipulation robots that we used for developing and testing our approaches.

classification techniques for supervised learning problems, and dimensionality reduction and clustering techniques for unsupervised learning problems. We discuss Bayesian model evaluation and model selection to choose between alternative models, and review graphical models as a tool to factorize large learning problems into feasible components.

In Chapter 3, we present a novel approach that enables a robot to learn the body schema of its manipulator from scratch using visual self-observation. In contrast to previous approaches, we estimate both the kinematic structure and the kinematic properties of the robot arm. We model the observations of each link of the arm as a Gaussian process and learn a Bayesian network that describes the kinematics of the whole system. The explicit representation of the kinematic structure allows the robot to detect and localize deviations between the model and the real arm to specific components of the network. Hence, the robot can efficiently adapt the model by re-learning only the mismatching parts. Our approach provides a flexible, probabilistic representation of robot kinematics and, furthermore, enables a manipulation robot to position its end effector accurately even in the presence of hardware failures and deformations. As a result, robots using our approach require less maintenance and can be used over longer periods of time without human intervention.

For many manipulation tasks in domestic environments, service robots need to open and close cabinets, for example, to stow or pick up objects. In Chapter 4, we show how our approach on body schema learning can be generalized to such articulated objects. We extend our approach by additional parametric models and use Bayesian model comparison to choose between the alternatives. This increases the robustness and efficiency of our approach while we keep the high flexibility of the Gaussian process models. In contrast to previous work, our approach applies to a significantly larger class of articulated objects and provides more accurate kinematic models. Furthermore, we can estimate the degrees of freedom of an articulated object and discover kinematic loops. We demonstrate that manipulation robots using our approach can learn accurate kinematic models of various doors and drawers, and operate them reliably. Complimentary to this, we demonstrate in Chapter 5 how a manipulation robot can recognize cabinet doors and drawers on dense depth images without requiring visual markers.

If a robot has tactile sensors in its gripper, it can use them to obtain additional information about the grasped objects. In Chapter 6 and 7, we present two novel approaches that manipulation robots can use to learn tactile object models. The first approach clusters the sensor data to create a tactile vocabulary. The robot uses this vocabulary to train a classifier which it can subsequently use to verify whether it has grasped the correct object. Our second approach is based on the temporal analysis of the sensor signal and allows a robot to recognize whether a grasped container is properly closed and whether it contains liquid. This ability is, for example, important for a domestic service robot that tidies up a table and needs to decide whether a juice bottle is full or empty and should be stored in the fridge or disposed in the trash can.

In Chapter 8, we show how a robot can learn novel manipulation tasks by observing a human instructor that repeatedly demonstrates this task. From these demonstrations, the robot extracts invariances in the execution of the task and infers from them a generalized task model. The robot can use this model to robustly reproduce the task even under different conditions. We model imitation learning as a probabilistic learning problem of a dynamic Bayesian network that encodes the task model using temporal and spatial object relations. In contrast to existing approaches, the factorized representation of the manipulation task as a dynamic Bayesian network allows us to dynamically add new constraints, for example, to avoid obstacles during reproduction, or to prefer a particular body posture. Our approach allows normal users to provide novel task descriptions to a manipulation robot in an intuitive way, which we consider an important prerequisite for the daily use of manipulation robots. Finally, we conclude this book with a summary of our results in Chapter 9 and give an outlook to future work.

To develop and test our approaches, we used different state-of-the-art mobile manipulators (see Figure 1.2): the first robot, Zora, is a B21R mobile base equipped with Schunk Powercube modules and tactile sensor arrays in the finger tips produced by Weiss robotics. The second robot, Cody, consists of a Segway mobile base, has two Meka arms, and is located at the Georgia Institute of Technology. The third robot, Marvin, is a PR2 robot from Willow Garage. It has a rich sensor suite including a tilting laser, two pairs of stereo cameras, a texture projector, and a tactile sensor array from PPS. By evaluating our approaches successfully on different experimental platforms, we ensure that our approaches also generalize to other mobile manipulation robots.

All of our approaches are based on state-of-the-art Bayesian learning techniques such as Gaussian processes, sample consensus methods, and graphical models. The probabilistic formulation of our approaches allows a robot to deal with uncertainties in the sensor observations and action execution and to consider them adequately during action planning. In an exhaustive set of experiments on real robots and in simulation we demonstrate that our approaches significantly reduce the dependency of manipulation robots on hand-crafted models and structured environments. Furthermore, we show that our approaches substantially increase the flexibility, adaptability and robustness of manipulation robots. We hope that our work contributes a small step to the goal of making natural environments accessible for mobile manipulation robots.

1.1 Key Contributions

The contributions presented in this book are innovative approaches that enable manipulation robots to learn probabilistic models of their sensors, actuators, and other objects in the world. We combine techniques from the fields of machine learning, artificial intelligence, and robotics with the goal to make manipulation robots more flexible, adaptive, and robust. In the following, we summarize our main contributions. This book presents:

- an approach for learning the body schema of a manipulation robot from scratch using visual self-observation, and a strategy to adapt it autonomously for tool use and in the case of hardware failures (Chapter 3),
- a general framework for learning kinematic models of articulated objects that allows robots to reliably operate doors and drawers (Chapter 4),
- a system to estimate such models from visual observation without requiring artificial markers (Chapter 5),
- a technique that enables a robot to use its tactile sensors to identify objects (Chapter 6),
- a method to estimate the internal state of the object being manipulated using tactile sensors (Chapter 7), and
- a solution to imitation learning with which a normal user can intuitively teach novel manipulation tasks that a robot can reproduce accordingly even under different conditions (Chapter 8).

1.2 Publications

This book is based on the doctoral thesis of Jürgen Sturm. Parts of this work have been published in international journals and collections, and presented at conferences and workshops.

PhD Thesis

- J. Sturm. Approaches to probabilistic model learning for mobile manipulation robots. *Freiburg University Library*, Freiburg i. Br., Germany, Sept. 2011.

Book Chapters

- J. Sturm, C. Plagemann, W. Burgard. Body schema learning. Chapter in *Towards Service Robots for Everyday Environments* Eds. E. Prassler et al., Springer, STAR 76, 130–161, Mar. 2012.

Journal Articles

- J. Sturm, C. Stachniss, and W. Burgard. A probabilistic framework for learning kinematic models of articulated objects. *Journal of Artificial Intelligence Research (JAIR)*. 41, July 2011.
- S. Chitta, J. Sturm, M. Piccoli, and W. Burgard. Tactile sensing for mobile manipulation. *IEEE Transactions on Robotics (T-RO)*. 27(3):558–568, June 2011.
- J. Sturm, C. Plagemann, and W. Burgard. Body schema learning for robotic manipulators from visual self-perception. *Journal of Physiology-Paris*, 103(3-5):220–231, Sept. 2009.

Conferences and Workshops

- T. Rühr, J. Sturm, D. Pangercic, M. Beetz, D. Cremers. A generalized framework for opening doors and drawers in kitchen environments. In *Proc. of the IEEE Intl. Conf. on Robotics and Automation (ICRA)*, St. Paul, MN, USA, May 2012.
- J. Becker, C. Bersch, D. Pangercic, B. Pitzer, T. Rühr, B. Sankaran, J. Sturm, C. Stachniss, M. Beetz, W. Burgard. Mobile manipulation of kitchen containers. In *Proc. of the Workshop on Workshop on Results, Challenges and Lessons Learned in Advancing Robots with a Common Platform at the Intl. Conf. on Intelligent Robot Systems (IROS)*, San Francisco, CA, USA, Sept. 2011.
- J. Hess, J. Sturm, W. Burgard. Learning the state transition model to efficiently clean surfaces with mobile manipulation robots. In *Proc. of the Workshop on Mobile Manipulation under Uncertainty at the IEEE Intl. Conf. on Robotics and Automation (ICRA)*, Shanghai, China, May 2011.
- J. Sturm, A. Jain, C. Stachniss, C.C. Kemp, and W. Burgard. Operating articulated objects based on experience. In *Proc. of the Intl. Conf. on Intelligent Robot Systems (IROS)*, Taipei, Taiwan, Oct. 2010.
- J. Sturm, K. Konolige, C. Stachniss, and W. Burgard. 3D pose estimation, tracking and model learning of articulated objects from dense depth

video using projected texture stereo. In *Proc. of the Workshop on Advanced Reasoning with Depth Cameras at the Robotics: Science and Systems Conf. (RSS)*, Zaragoza, Spain, Jun. 2010.

- J. Sturm, K. Konolige, C. Stachniss, and W. Burgard. Vision-based detection for learning articulation models of cabinet doors and drawers in household environments. In *Proc. of the IEEE Intl. Conf. on Robotics and Automation (ICRA)*, Anchorage, AK, May 2010.
- D. Meyer-Delius, J. Sturm, and W. Burgard. Regression-based online situation recognition for vehicular traffic scenarios. In *Proc. of the Intl. Conf. on Intelligent Robot Systems (IROS)*, St. Louis, MO, USA, Oct. 2009.
- A. Schneider, J. Sturm, C. Stachniss, M. Reisert, H. Burkhardt, and W. Burgard. Object identification with tactile sensors using bag-of-features. In *Proc. of the Intl. Conference on Intelligent Robot Systems (IROS)*, St. Louis, MO, USA, Oct. 2009.
- J. Sturm, V. Pradeep, C. Stachniss, C. Plagemann, K. Konolige, and W. Burgard. Learning kinematic models for articulated objects. In *Proc. of the Intl. Joint Conf. on Artificial Intelligence (IJCAI)*, Pasadena, CA, USA, Jul. 2009.
- H. Schulz, L. Ott, J. Sturm, and W. Burgard. Learning kinematics from direct self-observation using nearest-neighbor methods. In *Proc. of the German Workshop on Robotics*, Braunschweig, Germany, Jun. 2009.
- J. Sturm, C. Stachniss, V. Pradeep, C. Plagemann, K. Konolige, and W. Burgard. Towards understanding articulated objects. In *Proc. of the Workshop on Robot Manipulation at the Robotics: Science and Systems Conf. (RSS)*, Jun. 2009.
- C. Eppner, J. Sturm, M. Bennewitz, C. Stachniss, and W. Burgard. Imitation learning with generalized task descriptions. In *Proc. of the IEEE Intl. Conf. on Robotics and Automation (ICRA)*, Kobe, Japan, May 2009.
- J. Sturm, C. Stachniss, V. Pradeep, C. Plagemann, K. Konolige, and W. Burgard. Learning kinematic models for articulated objects. In *Proc. of the Learning Workshop (Snowbird)*, Clearwater, FL, USA, Apr. 2009.
- J. Sturm, C. Plagemann, and W. Burgard. Adaptive body scheme models for robust robotic manipulation. In *Robotics: Science and Systems Conf. (RSS)*, Zurich, Switzerland, Jun. 2008.
- J. Sturm, C. Plagemann, and W. Burgard. Body scheme learning and life-long adaptation for robotic manipulation. In *Proc. of the Workshop on Robot Manipulation at the Robotics: Science and Systems Conf. (RSS)*, Zurich, Switzerland, Jun. 2008.
- J. Sturm, C. Plagemann, and W. Burgard. Unsupervised body scheme learning through self-perception. In *Proc. of the IEEE Intl. Conf. on Robotics and Automation (ICRA)*, Pasadena, CA, USA, May 2008.

1.3 Additional Material

We provide additional material including videos, research papers, datasets, and errata. The website accompanying the book is located here:

`http://vision.in.tum.de/members/sturmju/book`

Furthermore, we released parts of our software as open-source to offer other researchers the opportunity to verify our results, evaluate our approaches on different data, and use our software in their research. In particular, we provide software implementing our body schema learning approach and the complete framework for learning kinematic models of articulated objects.

- The ZORA framework[1] implements our approach on body schema learning as described in Chapter 3. It is freely available under the GPL license. Furthermore, a detailed tutorial explains how to reproduce our results on various simulated manipulators.
- The ARTICULATION stack[2] provides several software libraries for learning kinematic models of articulated objects as described in Chapter 4 and Chapter 5. We released the software stack under the BSD license. Further, we provide several tutorials that explain in detail how kinematic models of articulated objects can be learned from observed trajectories and how the framework can be used with Python and C++.

1.4 Collaborations

Parts of this work have been developed in collaboration with other people. During my time as a PhD student, I supervised several master and bachelor projects. Chapter 6 on tactile object recognition is an extension of the bachelor thesis of Alexander Schneider (Schneider, 2009). Learning manipulation tasks from human demonstrations in Chapter 8 was originally addressed by Clemens Eppner in his master thesis (Eppner, 2008).

Furthermore, the approach on body schema learning as presented in Chapter 3 was jointly developed with Christian Plagemann. The operation of articulated objects as described in Chapter 4 was developed in close collaboration with Advait Jain and Charlie Kemp from the Georgia Institute of Technology. Finally, I closely collaborated with Sachin Chitta and Matthew Piccoli during my research stay at Willow Garage in 2009 during the development of our approach on tactile state recognition as described in Chapter 7.

[1] `http://www.informatik.uni-freiburg.de/~sturm/zora.html`

[2] `http://www.ros.org/wiki/articulation`

1.5 Symbols and Notation

Symbol	Meaning
x	scalar
$\mathbf{x}$	column vector
$\hat{\mathbf{x}}$	estimated value of $\mathbf{x}$
$\mathbf{x}^*$	optimal value of $\mathbf{x}$
A	matrix
$\mathrm{diag}(\mathbf{x})$	matrix with $\mathbf{x}$ on diagonal
$\lvert A \rvert$	determinant of A
$\{\ldots\}$	set
$(\ldots)$	vector
$\langle\ldots\rangle$	tuple (ordered set)
$\mathcal{D}$	training data
$\mathcal{M}$	model
$\mathbb{M}$	set of models
$p(\mathbf{x})$	probability distribution of a random variable $\mathbf{x}$
$p(\mathbf{y} \mid \mathbf{x})$	probability distribution of $\mathbf{y}$ conditioned on $\mathbf{x}$
$q(\mathbf{x})$	approximation of the probability distribution $p(\mathbf{x})$
$\mathcal{U}(S)$	uniform distribution over a set S
$\mathcal{N}(\boldsymbol{\mu}, \Sigma)$	normal distribution with mean $\boldsymbol{\mu}$ and covariance Σ
$\mathcal{N}(\mathbf{x}; \boldsymbol{\mu}, \Sigma)$	probability density at $\mathbf{x}$ of a normal distribution
$\mathbb{E}[\mathbf{x}]$	expected value of a random variable $\mathbf{x}$
$\nabla f(x)\big\rvert_{x=\hat{x}}$	gradient of $f(x)$ evaluated at $\hat{x}$
$\partial f / \partial x$	partial derivative of f with respect to x
$SO(3)$	Special orthogonal group of dimension 3. Elements of this group are 3D orientations.
$SE(3)$	Special Euclidean group of dimension 3. Elements of this group are 3D poses, consisting of 3D position and 3D orientation.

1.6 Acronyms

BFGS	Broyden-Fletcher-Goldfarb-Shanno method
BIC	Bayesian information criterion
BN	Bayesian network
CEP	Cartesian-space equilibrium point
DBN	dynamic Bayesian network
DH	Denavit-Hartenberg parameters
DMP	dynamic motion primitives
DOF	degrees of freedom
EM	expectation maximization
EPC	equilibrium point control
GP	Gaussian process
GN	Gauss-Newton algorithm
HMM	hidden Markov model
KLD	Kullback-Leibler divergence
LLE	locally linear embedding
LM	Levenberg-Marquardt algorithm
MAP	maximum a posteriori
ML	maximum likelihood
MLESAC	maximum likelihood consensus
MSAC	M-estimator sample consensus
PCA	principal component analysis
PROSAC	progressive sample consensus
RANSAC	random sample consensus
RBF	radial basis function
RL	reinforcement learning
RMS	root mean square error
ROS	robot operating system
SLAM	simultaneous localization and mapping

Chapter 2
Basics

The goal of this chapter is to provide the reader with an overview of the machine learning techniques used in this book. A good introduction to the field of machine learning in general can be found in the books of Bishop (2007) and MacKay (2003). This chapter starts with a review of common machine learning techniques for regression, classification, dimensionality reduction, and clustering problems. To compare and rank alternative models, we present in Section 2.2 several measures to evaluate the quality of a model and to select the best one. Finally, we introduce in Section 2.3 Bayesian networks as a tool to factorize high-dimensional learning problems into independent components.

2.1 Model Learning

One of the primary goals in the field of machine learning is to find a *model* that describes the dependency of one random variable from another one. In probability theory, this dependency is defined by the *conditional probability distribution*

$$p(\mathbf{y} \mid \mathbf{x}, \mathcal{M}) \tag{2.1}$$

which refers to the probability distribution of the random variable $\mathbf{y}$ given the value of the random variable $\mathbf{x}$ and a model $\mathcal{M}$. In the remainder of this chapter, we call $\mathbf{x} \in \mathcal{X}$ the *input variable* and $\mathbf{y} \in \mathcal{Y}$ the *target variable.* Correspondingly, we refer to the space of possible inputs as the *input space* $\mathcal{X}$ and to the space of possible targets as the *target space* $\mathcal{Y}$.

2.1.1 Regression

If the target space $\mathcal{Y}$ is continuous, the problem of estimating the conditional density function $p(\mathbf{y} \mid \mathbf{x})$ is called *regression.* If a deterministic relationship between input and target space exists, the model can be specified using a *regression function* that defines the functional mapping $f_{\mathcal{M}}$ from input to target space, i.e.,

$$\mathbf{y} = f_{\mathcal{M}}(\mathbf{x}). \tag{2.2}$$

J. Sturm: *Approaches to Probabilistic Model Learning*, STAR 89, pp. 13–33.
DOI: 10.1007/978-3-642-37160-8_2

In many practical problems, the observed targets are distorted by noise. This turns the problem of estimating $f_{\mathcal{M}}$ into a *noisy regression problem.* The relationship between inputs and targets thus becomes

$$\mathbf{y} = f_{\mathcal{M}}(\mathbf{x}) + \boldsymbol{\epsilon}, \tag{2.3}$$

where the regression function $f_{\mathcal{M}}(\mathbf{x})$ is given by the mean of the conditional probability distribution $p(\mathbf{y} \mid \mathbf{x})$ and the term $\boldsymbol{\epsilon}$ refers to additive noise which is typically assumed to be independent and identically distributed. The goal of regression is to estimate $f_{\mathcal{M}}(\mathbf{x})$ from a set of n observations $\mathcal{D} = \{(\mathbf{x}_i, \mathbf{y}_i)\}_{i=1}^{n}$. This set is also called the *training data* from which the model is estimated.

Parametric Regression

A parametric approach to regression is to express the unknown function with a function $f_{\mathcal{M},\boldsymbol{\theta}}$ that is parametrized by a vector $\boldsymbol{\theta}$. A simple model is the so-called *linear regression model* where the targets depend linearly on the inputs, i.e.,

$$\mathbf{y} = \boldsymbol{\theta}^T \mathbf{x} + \boldsymbol{\epsilon}. \tag{2.4}$$

The goal is then to select the parameter vector $\boldsymbol{\theta}$ that best fits the data, or equivalently, that maximizes the posterior probability after having observed the training data $\mathcal{D}$, i.e.,

$$\hat{\boldsymbol{\theta}} = \arg\max_{\boldsymbol{\theta}} p(\boldsymbol{\theta} \mid \mathcal{D}, \mathcal{M}). \tag{2.5}$$

By applying Bayes' rule and neglecting the prior probability of the training data $p(\mathcal{D})$ as it is independent of the choice of the parameter vector $\boldsymbol{\theta}$, we can rewrite this equation as

$$\hat{\boldsymbol{\theta}} = \arg\max_{\boldsymbol{\theta}} p(\mathcal{D} \mid \boldsymbol{\theta}, \mathcal{M})\, p(\boldsymbol{\theta} \mid \mathcal{M}), \tag{2.6}$$

where $p(\mathcal{D} \mid \boldsymbol{\theta}, \mathcal{M})$ is called the *data likelihood* and $p(\boldsymbol{\theta} \mid \mathcal{M})$ is the *prior* over the parameter space.

Linear Least Squares

For linear regression models as in Eq. (2.4) with normally distributed noise, Eq. (2.6) can be solved in closed form as a least squares problem. To see this, observe that each training sample yields one (linear) constraint of the following form, we obtain a linear system of equations:

$$\begin{aligned} \mathbf{y}_1 &= \mathbf{x}_1^T \boldsymbol{\theta}, \\ \mathbf{y}_2 &= \mathbf{x}_2^T \boldsymbol{\theta}, \\ &\vdots \\ \mathbf{y}_n &= \mathbf{x}_n^T \boldsymbol{\theta}. \end{aligned} \tag{2.7}$$

By stacking all target values in a vector $\mathbf{y} = (\mathbf{y}_1, \ldots, \mathbf{y}_n)^T$ and all inputs in a matrix $X = (\mathbf{x}_1, \ldots, \mathbf{x}_n)^T$, respectively, the above system of equations can be expressed conveniently by the equation

$$\mathbf{y} = X\boldsymbol{\theta}. \tag{2.8}$$

When the system is over-determined, i.e., we have more constraints than model parameters, Eq. (2.8) has typically no exact solution, which means that at least some of the residuals

$$\mathbf{r}_i := \mathbf{y}_i - \mathbf{x}_i^T \boldsymbol{\theta} \tag{2.9}$$

will be non-zero. Under the assumption that the targets $\mathbf{y}$ are perturbed by Gaussian noise, our goal is to maximize the data likelihood which is equivalent to minimizing the square of the residuals, i.e.,

$$\arg\min_{\boldsymbol{\theta}} \|\mathbf{y} - X\boldsymbol{\theta}\|^2. \tag{2.10}$$

To find this minimum, we expand the objective function

$$\|\mathbf{y} - X\boldsymbol{\theta}\|^2 = (\mathbf{y} - X\boldsymbol{\theta})^T(\mathbf{y} - X\boldsymbol{\theta}) \tag{2.11}$$

$$= \mathbf{y}^T\mathbf{y} - \mathbf{y}^T X\boldsymbol{\theta} - \boldsymbol{\theta}^T X^T \mathbf{y} + \boldsymbol{\theta}^T X^T X \boldsymbol{\theta} \tag{2.12}$$

and set the derivative to zero, i.e.,

$$\frac{\mathrm{d}\|\mathbf{y} - X\boldsymbol{\theta}\|^2}{\mathrm{d}\boldsymbol{\theta}} \stackrel{!}{=} 0 \tag{2.13}$$

$$\Rightarrow 2X^T X\boldsymbol{\theta} - 2X^T\mathbf{y} = 0 \tag{2.14}$$

$$\Rightarrow X^T X\boldsymbol{\theta} = X^T\mathbf{y} \tag{2.15}$$

This system can be efficiently solved for the model parameters $\boldsymbol{\theta}$ using Gauss elimination. Alternatively, the solution can be obtained by inverting $X^T X$ (e.g., using Cholesky decomposition) and left multiplying, i.e.,

$$\hat{\boldsymbol{\theta}} = (X^T X)^{-1} X^T \mathbf{y}. \tag{2.16}$$

The matrix $(X^T X)^{-1} X^T$ is called the pseudo-inverse of X.

Nonlinear Optimization

In the general case where $f_{\mathcal{M},\boldsymbol{\theta}}$ is any arbitrary function, no closed-form solution exists. In this case, iterative minimization techniques such as gradient descent, Gauss-Newton, or Levenberg-Marquardt can be used to find the best model parameter $\boldsymbol{\theta}$. Note that all of these methods only find a local minimum given an initial value for $\boldsymbol{\theta}$.

The most basic technique to nonlinear minimization is *gradient descent*. Sometimes, the method is also called steepest descent as it iteratively follows the gradient towards a local minimum of a function. In each iteration k, gradient descent adds small increments $\Delta\boldsymbol{\theta}$ to the current estimate $\boldsymbol{\theta}_k$ where the direction of the increment corresponds to the negative gradient of our objective function at the current location, i.e.,

$$\Delta\boldsymbol{\theta} = -\gamma\nabla\sum_i(\mathbf{y}_i - f_{\mathcal{M},\boldsymbol{\theta}_k}(\mathbf{x}_i))^2, \tag{2.17}$$

where ∇ refers to the gradient operator. γ denotes the step size and has to be chosen carefully, e.g., via a line search. The increment can be added to the current estimate $\boldsymbol{\theta}_k$ to obtain a new estimate $\boldsymbol{\theta}_{k+1}$ with

$$\boldsymbol{\theta}_{k+1} = \boldsymbol{\theta}_k + \Delta\boldsymbol{\theta}. \tag{2.18}$$

This procedure is iterated with the new estimate $\boldsymbol{\theta}_{k+1}$ until a convergence criteria is met, e.g., the magnitude of the residual or the increment falls below a certain threshold, or the maximum number of iterations is exceeded. While gradient descent can be applied to a large class of nonlinear minimization problems, other methods are typically preferred as they converge significantly faster.

The *Gauss-Newton algorithm* approximates the (nonlinear) regression model of Eq. (2.3) by its first order Taylor expansion in the parameter vector $\boldsymbol{\theta}$ evaluated at the current estimate $\boldsymbol{\theta}_k$, i.e.,

$$\begin{aligned} f_{\mathcal{M},\boldsymbol{\theta}}(\mathbf{x}_i) &= f_{\mathcal{M},\boldsymbol{\theta}_k+\Delta\boldsymbol{\theta}}(\mathbf{x}_i) && (2.19)\\ &\approx f_{\mathcal{M},\boldsymbol{\theta}_k}(\mathbf{x}_i) + \underbrace{\left.\frac{\mathrm{d}f_{\mathcal{M},\boldsymbol{\theta}}(\mathbf{x}_i)}{\mathrm{d}\boldsymbol{\theta}}\right|_{\boldsymbol{\theta}=\boldsymbol{\theta}_k}}_{=:J_i}\Delta\boldsymbol{\theta} && (2.20)\\ &= f_{\mathcal{M},\boldsymbol{\theta}_k}(\mathbf{x}_i) + J_i\Delta\boldsymbol{\theta}, && (2.21) \end{aligned}$$

where J_i is the derivative of $f_{\mathcal{M},\boldsymbol{\theta}}$ with respect to $\boldsymbol{\theta}$ evaluated at $\boldsymbol{\theta}_k$ for $\mathbf{x}_i$. With this approximation, we obtain an expression that is linear in its parameters. Again, our goal is to minimize the squared residuals, i.e.,

$$\arg\min_{\boldsymbol{\theta}}\sum_i(\mathbf{y}_i - f_{\mathcal{M},\boldsymbol{\theta}_k}(\mathbf{x}_i) - J_i\Delta\boldsymbol{\theta})^2. \tag{2.22}$$

In matrix notation, this corresponds to

$$\arg\min_{\boldsymbol{\theta}}\|\mathbf{Y} - f_{\mathcal{M},\boldsymbol{\theta}_k}(X) - J\Delta\boldsymbol{\theta}\|^2, \tag{2.23}$$

where J is the Jacobian of $f_{\mathcal{M},\boldsymbol{\theta}}$ whose i-th row equals J_i. After expanding this and setting the derivative to zero, we obtain

$$J^TJ\Delta\boldsymbol{\theta} = J^T(\mathbf{Y} - f_{\mathcal{M},\boldsymbol{\theta}_k}(X)). \tag{2.24}$$

This equation can be solved by Gaussian elimination or matrix inversion as detailed in the previous subsection to obtain the parameter increment

$\Delta\boldsymbol{\theta}$. Typically, the Gauss-Newton algorithm converges very quickly (i.e., quadratically) under the assumption that we have a good initial estimate $\boldsymbol{\theta}_0$ and that the regression model is sufficiently well approximated by the linearization.

An extension of this method is the *Levenberg-Marquardt (LM) algorithm.* It interpolates between the Gauss-Newton algorithm and simple gradient descent by adding a damping factor to the Jacobian in Eq. (2.24), i.e.,

$$(J^T J + \lambda I)\Delta\boldsymbol{\theta} = J^T(\mathbf{Y} - f_{\mathcal{M},\boldsymbol{\theta}_k}(X)), \tag{2.25}$$

where I is the identity matrix and $\lambda \in \mathbb{R}^+$ is the so-called damping factor. When a small damping factor is chosen, Levenberg-Marquardt effectively reduces to Gauss-Newton, while a large damping factor leads to increments similar to gradient descent.

Note that in contrast to the linear case, all of these techniques are prone to local optima. This is in particular a problem when the training data – in addition to the Gaussian noise – is contaminated with outliers. In this case, re-starts of the minimization with random initializations can help to find the global optimum.

Sampling Consensus Methods

If the parameter space is large, random search alone is not very effective to escape from local minima. *Sampling consensus methods* have been proven to be very useful in this context and are widely used in practice. The general idea behind these methods is that the training data comes from a mixture distribution composed of inliers and outliers. The inliers are noisy, Gaussian observations of the true underlying model, while the outliers originate from a completely unrelated (and generally unknown) distribution. As the goal is to estimate the model parameters from the inlier distribution, the random sample consensus algorithm (RANSAC) repeatedly samples minimal subsets of observations from the training data and uses them to estimate an initial guess for the parameter vector (Fischler and Bolles, 1981). Based on this estimate, the training data is split into inliers and outliers. Finally, only the parameter vector with the largest number of inliers is kept and optimized further over the inlier set. Torr and Zisserman (2000) generalized this algorithm by assigning a constant penalty to all training samples above a particular error threshold using a truncated error function. As a consequence, M-estimator sample consensus (MSAC) can re-assign outliers to inliers and vice versa during optimization of the model parameters. Both RANSAC and MSAC require a fixed threshold to determine whether a sample belongs to the inlier or outlier set. Torr and Zisserman (2000) showed that this threshold (or mixing ratio between inliers and outliers) can be estimated alongside during parameter optimization using the expectation minimization (EM) algorithm with only a minimal computational overhead. As this approach maximizes the likelihood of all observations, it is called maximum likelihood sample consensus

(MLESAC). When the training samples have associated confidence values, it has been shown that progressive sample consensus (PROSAC) can be used to guide sampling and speed up convergence significantly (Chum and Matas, 2005).

Parametric regression is probably the most widely used method of regression. In this book, we use parametric models in combination with sampling consensus methods and nonlinear optimization to learn models for the motion of articulated objects from noisy observations in Chapter 4.

Nonparametric Regression

A different class of regression techniques are so-called *nonparametric methods*. These approaches do not require an explicit parametrization of the underlying function $f_{\mathcal{M}}$. Instead, the regression function $f_{\mathcal{M}}$ is implicitly defined by the training data. As a consequence, nonparametric methods are not limited to a particular function form. However, they require the whole training data for making predictions such that essentially all training samples can be considered as parameters of the model.

The *Gaussian process (GP) model* assumes that the training samples in $\mathcal{D}$ are samples of a joint Gaussian distribution (Rasmussen and Williams, 2006), i.e.,

$$\mathbf{y} \sim \mathcal{N}(\boldsymbol{\mu}, K), \tag{2.26}$$

where $\mathbf{y} = (y_1, \ldots, y_n)^T$ is the vector of the observed, one-dimensional target values. Without loss of generality, we assume that the mean of this distribution is zero, i.e., $\boldsymbol{\mu} = \mathbf{0}$. In case that a different mean function m is desired, it can be subtracted from the observations to create a zero-mean Gaussian process, i.e., by setting $\overline{y}_i := y_i - m(\mathbf{x}_i)$.

The interesting part of the GP model is the covariance matrix K. It is typically specified using a *covariance function* $k(\mathbf{x}_i, \mathbf{x}_j)$, i.e.,

$$K_{ij} := \operatorname{cov}(y_i, y_j) = k(\mathbf{x}_i, \mathbf{x}_j). \tag{2.27}$$

This function defines the covariance between any two targets y_i and y_j given their input vectors $\mathbf{x}_i$ and $\mathbf{x}_j$ as parameters. A popular choice is the squared exponential covariance function,

$$k_{SE}(\mathbf{x}, \mathbf{x}'; \boldsymbol{\theta}) = \sigma_f^2 \exp\left(-\frac{1}{2}\sum_{i=1}^{D}\frac{(x_i - x_i')^2}{l_i^2}\right) + \sigma_n^2 \delta_{ij}, \tag{2.28}$$

where $\boldsymbol{\theta} = (\sigma_f, \sigma_n, l_1, \ldots, l_D)$ are the so-called *hyper-parameters* and D gives the number of input dimensions. The amplitude σ_f defines the scale of the covariances between data points, the process noise σ_n describes the amount of noise expected in the training data, and the length-scale parameters $l_1, \ldots, l_D$ define the smoothness of the function in each input dimension. Similar to

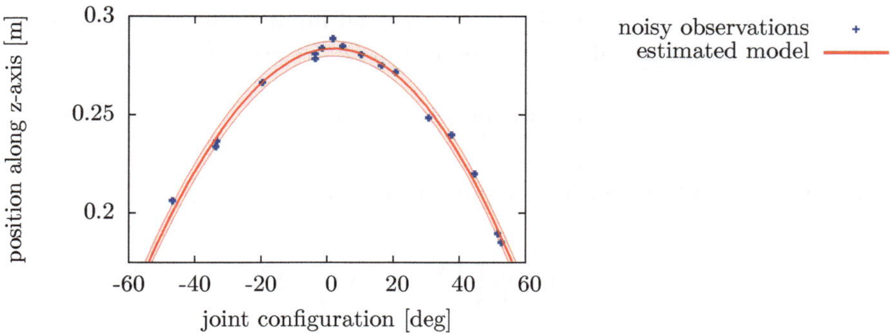

Fig. 2.1 Example of a regression problem. Here, the robot learns a model of its arm kinematics from real data using a Gaussian process. The red line corresponds to the mean prediction of the learned model, the shaded area to the standard deviation of the predicted variance.

parametric regression, these hyper-parameters can be learned from training data by maximizing the marginal data likelihood.

Figure 2.1 shows an example of Gaussian process regression that was learned on real observations from Chapter 3. Here, the robot observed the position of its end effector in different joint configurations, and learned a GP model to the arm kinematics. We also employ GPs in Chapter 4 for learning nonparametric link models of articulated objects, for example to describe the motion of a garage door.

2.1.2 Classification

When the target space is finite, learning the conditional probability distribution $p(\mathbf{y} \mid \mathbf{x}, \mathcal{M})$ is called a *classification* problem. Classification is used to assign a novel input vector $\mathbf{x}$ to one of k classes $\mathcal{Y} = \{1, \ldots, k\}$ by assigning it to the most likely class, i.e.,

$$\hat{\mathbf{y}} = \arg\max_{\mathbf{y} \in \mathcal{Y}} p(\mathbf{y} \mid \mathbf{x}, \mathcal{M}). \tag{2.29}$$

In the following, we briefly review three classification techniques, called the *naive Bayes classifier*, the *bag-of-features* approach, and *decision tree learning.*

Naive Bayes

The naive Bayes classifier (Duda et al., 1973) is a simple approach to classification. It is based on the strong assumption that all input dimensions of $\mathbf{x} \in \mathbb{R}^D$ are mutually independent of each other given the class label, i.e.,

$$p(\mathbf{x} \mid \mathbf{y}) = \prod_{i=1}^{D} p(x_i \mid \mathbf{y}), \tag{2.30}$$

where $\mathbf{x} = (x_1, \ldots, x_D)^T$ is the D-dimensional input vector. With this, we can rewrite the classification function of Eq. (2.29) using Bayes' rule, i.e.,

$$\hat{\mathbf{y}} = \arg\max_{\mathbf{y}\in\mathcal{Y}} p(\mathbf{x} \mid \mathbf{y})p(\mathbf{y}) \tag{2.31}$$

and apply the independence assumption which gives us

$$\hat{\mathbf{y}} = \arg\max_{\mathbf{y}\in\mathcal{Y}} \left(\prod_{i=1}^{D} p(x_i \mid \mathbf{y})\right) p(\mathbf{y}). \tag{2.32}$$

This factorization simplifies the learning problem significantly: instead of having to learn the joint probability distribution $p(\mathbf{y} \mid \mathbf{x})$, the individual classification models $p(x_i \mid \mathbf{y})$ can be learned separately from the data. Depending on the underlying classification models, these learning step can be implemented efficiently, for example, when histogram models are being used. In spite of the strong independence assumptions between the input dimensions, naive Bayes has been shown to perform quite well on a large number of complex real-world problems.

Bag-of-Features

The naive Bayes classifier forms the basis of the bag-of-features approach which was originally developed for object classification tasks in computer vision (Lewis, 1998; Fei-Fei and Perona, 2005). Instead of operating directly on the pixels of an image, the bag-of-features approach extracts an intermediate set of features from the images and learns the classification model only on these features. By counting how often a particular feature $\mathbf{x}$ is present in an image I, one obtains histogram distributions $p(\mathbf{x} \mid I)$ of features in the image. In the training phase, a *codebook* C of these histogram distributions $p(\mathbf{x} \mid \mathbf{y})$ is learned that expresses the probabilistic relationship between features and object classes. For classifying a novel image, the feature histogram can be computed and compared to the stored histograms in the codebook. Popular histogram distance metrics include χ^2, Kullback-Leibler divergence or histogram intersection. The bag-of-features classifier then assigns the class to an object that maximizes the likelihood

$$\hat{\mathbf{y}} = \arg\min_{\mathbf{y}\in\mathcal{Y}} p\Big(d\big(p(\mathbf{x} \mid I), p\left(\mathbf{x} \mid \mathbf{y}, C\right)\big)\Big)p(\mathbf{y}), \tag{2.33}$$

where $d(\cdot,\cdot)$ refers to the chosen distance metric between histograms. The visual vocabulary required to extract features from an image can automatically be constructed using unsupervised clustering techniques such as k-means as discussed later in this chapter. In this book, we apply the bag-of-features approach in Chapter 6 to recognize objects in tactile sensor images.

Decision Tree Learning

Decision tree learning is an approach to infer a set of rules from the training data in order to predict the target class. One algorithm for learning decision trees efficiently from training data is the so-called C4.5 algorithm (Quinlan, 1993). The learning procedure of C4.5 starts by selecting an attribute that most effectively splits the data in the target classes based on entropy reduction. The *entropy* H of a training set $\mathcal{D}$ with respect to the target classes is defined as

$$H(\mathcal{D}) := -\sum_{\mathbf{y}\in\mathcal{Y}} p(\mathbf{y} \mid \mathcal{D}) \log p(\mathbf{y} \mid \mathcal{D}), \tag{2.34}$$

where $p(\mathbf{y} \mid \mathcal{D})$ is the occurrence probability of the class $\mathbf{y}$ in the training data $\mathcal{D}$. A split s is defined by a split value $s_{\text{value}} \in \mathbb{R}$ in a particular input dimension (or *attribute*) $s_{\text{attr}} \in \{1, \dots, D\}$. A split divides the training data $\mathcal{D}$ into two subsets

$$\mathcal{D}_{\leq} := \{(\mathbf{x}, \mathbf{y}) \in \mathcal{D} \mid x_{s_{\text{attr}}} \leq s_{\text{value}}\}, \tag{2.35}$$

$$\mathcal{D}_{>} := \{(\mathbf{x}, \mathbf{y}) \in \mathcal{D} \mid x_{s_{\text{attr}}} > s_{\text{value}}\}. \tag{2.36}$$

From all possible splits, C4.5 now selects the one with the highest *information gain*, i.e.,

$$s^* = \arg\max_{s} I(\mathcal{D}; s), \tag{2.37}$$

where the information gain is defined as the reduction in the entropy of the resulting sets compared with the initial set:

$$I(\mathcal{D}; s) := H(\mathcal{D}) - H(\mathcal{D} \mid s). \tag{2.38}$$

The *conditional entropy* $H(\mathcal{D} \mid s)$ is defined as

$$\begin{aligned} H(\mathcal{D} \mid s) := {} & H(\mathcal{D}_{\leq}) p(x_{s_{\text{attr}}} \leq s_{\text{value}} \mid (\mathbf{x}, \mathbf{y}) \in \mathcal{D}) + \\ & H(\mathcal{D}_{>}) p(x_{s_{\text{attr}}} > s_{\text{value}} \mid (\mathbf{x}, \mathbf{y}) \in \mathcal{D}). \end{aligned} \tag{2.39}$$

Each split s corresponds to a node of the decision tree with two children. The same procedure is then repeated for the resulting subsets $\mathcal{D}_{\leq}$ and $\mathcal{D}_{>}$, until the leaves are homogeneous with respect to the target class, i.e., the entropy in the dataset of the leaf with respect to the target classes is zero. An important step after training is pruning to avoid over-fitting to the training data. This is done by replacing a whole subtree by a leaf node if the expected error rate (computed on a separate test dataset held out during training) in the subtree is greater than in the single leaf.

In this book, we use decision tree learning for the classification of the state of objects based on tactile sensor observations in Chapter 7. A simple example of a binary classification problem is depicted in Figure 2.2. Here, a manipulation robot learns to discriminate empty from full bottles using tactile sensing. In this case, the inputs correspond to tactile features that are extracted from the high-frequency components of the sensor signal.

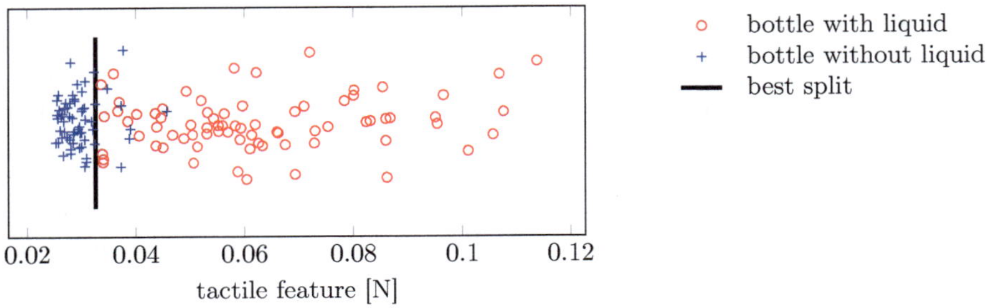

Fig. 2.2 Example of a binary classification problem where the robot learns a decision tree to discriminate full from empty bottles using tactile sensing. The one-dimensional tactile features are spread out on the y-axis to improve the readability of the plot.

2.1.3 Dimensionality Reduction

So far, we have assumed that both the inputs as well as the targets are fully *observable*, i.e., that the training set contains both input and target vectors. The previously discussed regression and classification techniques therefore fall in the category of *supervised learning* approaches. In contrast, *unsupervised learning* refers to the class of problems where the values of the target variables $\mathbf{y}$ are unknown. A random variable that is not observable is also called *latent*, and so the target space is also called the *latent space*. The goal in unsupervised learning is to reveal the structure underlying the training data in the input space and to establish a suitable mapping to the latent space.

Principal Component Analysis

Principal component analysis (PCA) is a classical tool in statistics to reveal the main axes of variation in a data set. The goal of PCA is to transform a set of possibly correlated variables into a set of uncorrelated variables using an orthogonal projection. PCA assumes that the data is centered around the origin. The covariance of the data can be estimated using

$$\mathrm{Cov} := \frac{1}{n-1} \sum_{i=1,\ldots,n} \overline{\mathbf{x}}_i \overline{\mathbf{x}}_i^T, \tag{2.40}$$

where n is the number of data samples in the training set, and $\overline{\mathbf{x}}_i$ are the zero-centered data samples, i.e., $\overline{\mathbf{x}}_i := \mathbf{x}_i - \frac{1}{n}\sum_{j=1}^{n} \mathbf{x}_j$. Assuming that the covariance matrix is positive-definite, it can be decomposed in its eigenvalues and eigenvectors, i.e.,

$$\mathrm{Cov} = W \Sigma W^T, \tag{2.41}$$

where $\Sigma = diag(\sigma_1^2, \ldots, \sigma_D^2)$ is the diagonal matrix of the squared eigenvalues, $W = (w_1, \ldots, w_D)$ are the corresponding eigenvectors, and D is the number of input dimensions. The number of dimensions of the resulting latent variable can be reduced while maximizing the variance by keeping only the d dimensions with the largest variance. The resulting projection becomes

$$\mathbf{y} = W_d^T \mathbf{x}, \tag{2.42}$$

where $W_d = (\mathbf{w}_{i_1}, \ldots, \mathbf{w}_{i_d})$ are the $d < D$ eigenvectors associated with the d largest eigenvalues. Different methods exist for choosing the number of dimensions d of the latent space. Depending on the application, d has a fixed value (for example, $d = 2$ for visualizing high dimensional data), or it is chosen in such a way that a particular percentage, for example, 95 %, of the original variance is preserved. As PCA provides a linear mapping from the data space to the latent space, the dimensionality reduction based on PCA works well if the input data lies on or close to a linear manifold. However, if the data lies on nonlinear manifolds, it can't be characterized well with a single linear projection.

Locally Linear Embedding

This problem is addressed by *Locally Linear Embedding (LLE)* which is a nonlinear method for dimensionality reduction. The goal of LLE is to find a mapping from input space to latent space that locally preserves the distances of neighboring data points both in the input and the latent space (Roweis and Saul, 2000). While the exact derivation of LLE is somewhat involved, the general idea of LLE can intuitively be described as follows. First, LLE finds for each data sample i the k-nearest neighbors of the input $\mathbf{x}_i$, denoted as the sequence $\langle \mathbf{x}_{i_1}, \ldots, \mathbf{x}_{i_k} \rangle$. Second, it seeks a weight vector $\mathbf{w}_i$ for each data sample that can be used to reconstruct each input vector $\mathbf{x}_i$ as a linear combination of its neighbors, i.e.,

$$\mathbf{w}_i = \arg \min_{\substack{\mathbf{w} \in \mathbb{R}^D, \\ \|\mathbf{w}\|=1}} \left(\mathbf{x}_i - \sum_{j \in \{1,\ldots,k\}} w_{i_j} \mathbf{x}_{i_j} \right), \tag{2.43}$$

where $\mathbf{x}_{i_j}$ denotes the j-th nearest neighbor of the i-th data sample and $\mathbf{w}_i = (w_{i_1}, \ldots, w_{i_k})$ correspondingly refers to the i-th weight vector. Under the constraint that each weight vector has unit length, i.e., $\|\mathbf{w}\| = 1$, this minimization can be solved in closed form. In the third step, LLE seeks for latent vectors $\mathbf{y}_i$ such that the same reconstruction property also holds in the latent space, i.e.,

$$\mathbf{y}_i = \arg \min_{\mathbf{y}} \left(\mathbf{y}_i - \sum_{j \in \{1,\ldots,k\}} w_{i_j} \mathbf{y}_{i_j} \right), \tag{2.44}$$

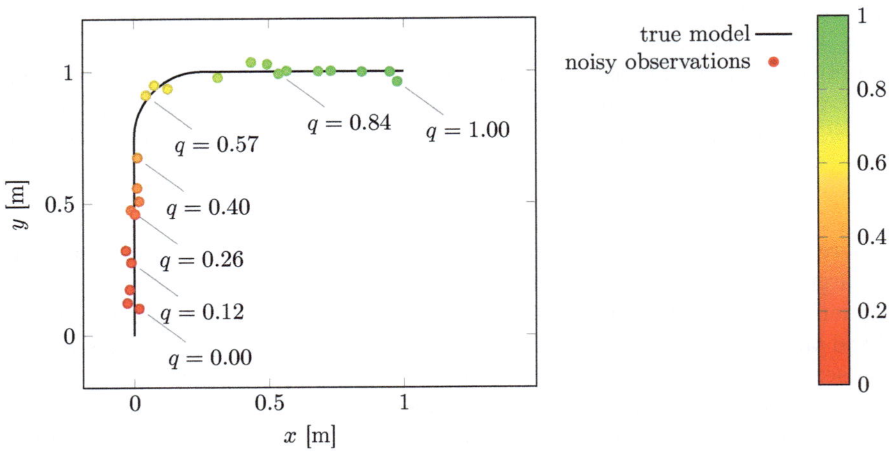

Fig. 2.3 Example of a dimensionality reduction problem. Here, LLE is used to infer the latent configuration q from pose observations (x, y) of a rolling garage door. The inferred configurations are encoded by the color.

where $\mathbf{y}_{i_j}$ refers to the j-th nearest neighbor of the latent coordinates corresponding to the i-th data sample. With a few additional constraints, the minimization in Eq. (2.44) can be solved again in closed form. The advantage of LLE over PCA is that it can recover both linear and nonlinear manifolds from the training data. The downside is that LLE is more sensitive to noise and computationally more demanding.

We use both PCA and LLE to infer the latent configuration of articulated objects in Chapter 4. Figure 2.3 shows an example where LLE has been applied to a sequence of pose observations of a rolling garage door. The robot observes the poses in 3D space and uses LLE to estimate their corresponding latent, one-dimensional configurations. The circles in the figure correspond to the observations of the xy-position, while the color of the circles encodes the latent configuration as recovered by LLE, i.e., the one-dimensional configuration of the garage door.

2.1.4 Clustering

The goal of clustering it to assign each training sample to one of k different clusters, i.e., it assumes a finite latent space with $\mathcal{Y} = \{1, \ldots, k\}$. This means that one needs to find a mapping from input vectors to clusters that minimizes the distances between input vectors belonging to the same cluster and maximizes the distances between input vectors from different clusters. In this section, we introduce *k-means clustering* and the *expectation maximization algorithm* as two popular clustering techniques.

K-means

K-means defines a cluster $i = 1, \ldots, k$ by its center (or mean) $\boldsymbol{\mu}_i \in \mathcal{X}$ in input space. Initially, these cluster centers are initialized to random samples from the training set. In each iteration, each training sample is assigned to the nearest cluster, i.e., the partition for cluster j becomes

$$S_i \leftarrow \{\mathbf{x} \in \mathcal{D} \mid d(\mathbf{x}, \boldsymbol{\mu}_i) \leq d(\mathbf{x}, \boldsymbol{\mu}_{i'}) \text{ for all } i' = 1, \ldots, k\}. \tag{2.45}$$

Subsequently, the cluster centers are re-assigned to the mean of all assigned input vectors, i.e.,

$$\boldsymbol{\mu}_i \leftarrow \frac{1}{|S_i|} \sum_{\mathbf{x} \in S_i} \mathbf{x}. \tag{2.46}$$

This procedure is repeated until the cluster centers have converged. In this book, we apply the k-means algorithm to generate a feature vocabulary from tactile images in Chapter 6 for object recognition using the bag-of-features approach introduced earlier in this chapter.

Expectation Maximization

Expectation maximization (EM) is a general method to find maximum likelihood estimates of parameters in latent variable models (Dempster et al., 1977; McLachlan and Krishnan, 1997). EM is a generalization of the k-means algorithm and similar to k-means, it iteratively estimates the values of the latent variables and re-estimates the model parameters given these values.

One particular application of EM is to estimate the parameters of a Gaussian mixture distribution given by

$$p(\mathbf{x}) = \sum_{i=1,\ldots,k} \pi_i \mathcal{N}(\mathbf{x} \mid \boldsymbol{\mu}_i, \Sigma_i), \tag{2.47}$$

where π_i are the so-called mixing coefficients (all positive and summing to one), and $\mathcal{N}(\mathbf{x} \mid \boldsymbol{\mu}_i, \Sigma_i)$ are the individual Gaussian mixture components. The goal is to infer the assignment of training samples to mixture components and to their parameters $\boldsymbol{\mu}_i, \Sigma_i$. The EM method solves this estimation problem iteratively in two steps. In the *expectation step*, the responsibilities of each mixture component for each data sample are computed based on the current estimate of the model parameters, i.e., for all mixture components i and all data samples j

$$\gamma_{ij} \leftarrow \frac{\pi_i \mathcal{N}(\mathbf{x} \mid \boldsymbol{\mu}_i, \Sigma_i)}{\sum_{l=1,\ldots,k} \pi_l \mathcal{N}(\mathbf{x} \mid \boldsymbol{\mu}_l, \Sigma_l)}. \tag{2.48}$$

Subsequently, in the *maximization step*, the model parameters are re-estimated given the current assignments of training samples to mixture components, i.e., for all i we set

$$\boldsymbol{\mu}_i \leftarrow \frac{\sum_{j=1,\ldots,n} \gamma_{ij} \mathbf{x}_j}{\sum_{j=1,\ldots,n} \gamma_{ij}}, \tag{2.49}$$

$$\Sigma_i \leftarrow \frac{\sum_{j=1,\ldots,n} \gamma_{ij} (\mathbf{x}_j - \boldsymbol{\mu}_i)(\mathbf{x}_j - \boldsymbol{\mu}_i)^T}{\sum_{j=1,\ldots,n} \gamma_{ij}}, \tag{2.50}$$

and

$$\pi_i \leftarrow \frac{1}{n} \sum_{j=1,\ldots,n} \gamma_{ij}. \tag{2.51}$$

This process is repeated until the parameters converge. As both EM and k-means are greedy methods, they are not guaranteed to find the global optimum. A practical solution is to re-start the algorithm several times with randomized initializations and to select the solution with the highest data likelihood.

In Chapter 4, we apply the EM algorithm to estimate both the model parameters and the outlier assignments from a sequence of pose observations in a Gaussian mixture model. Figure 2.4 shows an example problem from this chapter where a robot has observed a sequence of noisy poses of a cabinet door and aims to fit a circular model. Some of the observations, however, are real outliers that need to be detected and excluded in the estimation of the arc. In the figure, the colors of the dots encode the estimated probability of a sample being an inlier and the black line corresponds to the circular model fitted to the inliers.

2.2 Model Comparison and Model Selection

In the previous section, we introduced various methods for learning models from noisy data. After several alternative models have been learned from the data, the question arises how these models can be compared to each other. In this section, we present different approaches to evaluate and rank alternative models. When evaluated on independent and identically distributed data sets, the *root mean square error* or the *data likelihood* are suitable measures to compare alternative models even of different complexity. However, as additional data is typically either not available or costly to acquire, *cross-validation* techniques offer a solution where the limited training data can both be used for learning and validation. Finally, Bayesian model comparison allows to directly compare alternative models by their *posterior probability*, but requires a suitable prior over the model space.

2.2.1 Root Mean Square Error

The root mean square (RMS) error measures the derivation of the model predictions from the observations. It is computed as the root of the average of

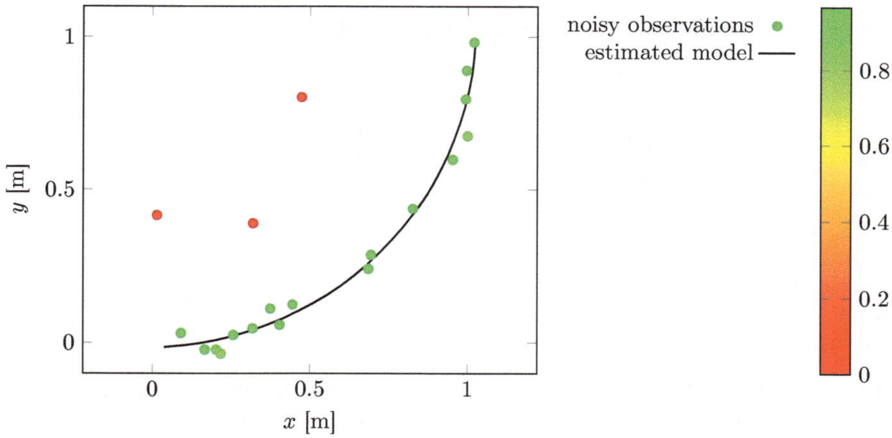

Fig. 2.4 Example of a clustering problem. EM iteratively assigns observations to the inlier and outlier component, and estimates the parameters of the circular model from the inliers. The color encodes the assignment of each observation to the mixture components.

the squared differences between predictions and observations, and is defined as

$$E_{\mathrm{RMS}}(\mathcal{D} \mid \mathcal{M}, \boldsymbol{\theta}) = \sqrt{\frac{1}{n} \sum_{i=1,\ldots,n} \| f_{\mathcal{M}}(\mathbf{x}_i; \boldsymbol{\theta}) - \mathbf{y}_i \|^2}, \tag{2.52}$$

where $\| \cdot \|$ is a distance metric in the target space, $\mathcal{M}$ is the model, and $\boldsymbol{\theta}$ its parameter vector. The division by n allows to compare data sets of different sizes. The advantage of the RMS error is that it can be intuitively interpreted, as it is measured in the same units as the target variable.

2.2.2 Data Likelihood

The data likelihood can also be used to measure the quality of a model. It is defined as the product over all observation likelihoods of an evaluation set $\mathcal{D} = \{(\mathbf{x}_i, \mathbf{y}_i)\}_{i=1}^{n}$, i.e.,

$$p(\mathcal{D} \mid \mathcal{M}, \boldsymbol{\theta}) = \prod_{i=1,\ldots,n} p(\mathbf{y}_i \mid \mathbf{x}_i) \tag{2.53}$$

$$\propto \prod_{i=1,\ldots,n} \exp\left(-(f_{\mathcal{M}}(\mathbf{x}_i; \boldsymbol{\theta}) - \mathbf{y}_i)^T \Sigma^{-1} (f_{\mathcal{M}}(\mathbf{x}_i; \boldsymbol{\theta}) - \mathbf{y}_i)\right), \tag{2.54}$$

where the second line holds under a Gaussian noise assumption with zero mean and covariance Σ.

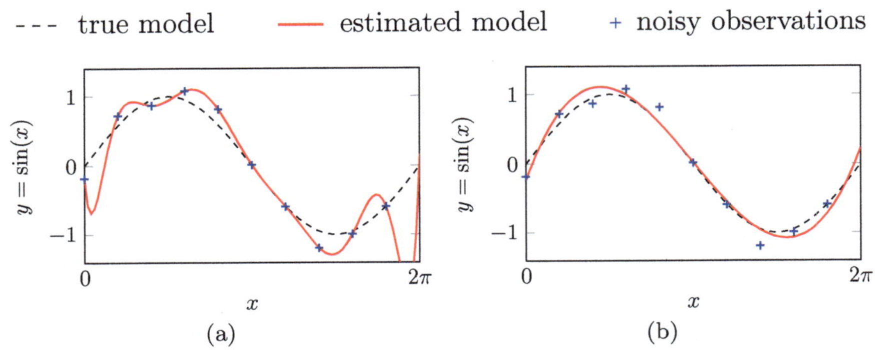

Fig. 2.5 These figures illustrate the problem of over-fitting. (a) The model fits exactly to the training points but generalizes poorly to previously unseen data. (b) An alternate model matches the training data less accurately, but generalizes better to new data.

2.2.3 Cross-Validation

Note that evaluating the RMS error or the data likelihood on the training set alone does not provide a good indicator of the predictive accuracy of a model on new data. The reason for this is that a complex model can fit the training data very accurately but generalize poorly to new data. This effect is called *over-fitting*, and is visualized in Figure 2.5. Although the RMS error of the model in Figure 2.5a is close to zero on the training set, it does not generalize well to new data. In contrast, the model in Figure 2.5b has a slightly higher RMS error on the training set, but generalizes much better to previously unseen data. A solution to this problem is to evaluate the data likelihood on an independent set of validation data that is drawn from the same distribution as the training data. In practice, however, the amount of available training data is strongly limited. Therefore, the expected *predictive error* of a model has to be estimated as efficiently as possible from the limited training data.

Cross-validation solves this problem by splitting the available data into mutually exclusive training and validation sets. Typically, several rounds of cross-validation are carried out to average over the expected predictive error. Different strategies for drawing the training and validation sets from the data exist. The most popular strategy is called *k-folds cross-validation* where in each of the k rounds a proportion $1/k$ of the data is left out for assessing the performance. *Leave-one-out cross-validation* is the extreme case where $k = n$, i.e., in each of the n rounds the model is trained from all but one sample, and evaluated on the remaining one. Cross-validation comes at an additional computational cost as the model needs to be learned anew in each of the k rounds. If there are several regularization parameters that need to be evaluated, the number of rounds can grow exponentially with the number of parameters (Bishop, 2007). Therefore, various other techniques have been developed that estimate the quality of a model directly from the data by incorporating the model complexity as a penalty term in the computation.

2.2.4 Bayesian Model Comparison

The Bayesian solution to the problem of model comparison is to assign a probability $p(\mathcal{M} \mid \mathcal{D})$ to each model that reflects the uncertainty in model choice (MacKay, 2003). From a Bayesian perspective, it is not even necessary to choose a single model, but the law of total probability can be applied to average over all models, i.e.,

$$p(\mathbf{y} \mid \mathbf{x}, \mathcal{D}) = \sum_{\mathcal{M} \in \mathbb{M}} p(\mathbf{y} \mid \mathbf{x}, \mathcal{D}, \mathcal{M}) p(\mathcal{M} \mid \mathcal{D}). \tag{2.55}$$

In practice, reasoning with all possible models quickly becomes intractable and therefore, often only the single most-likely model $\hat{\mathcal{M}}$ is considered. This is reasonable when the likelihood for the most-likely model dominates the likelihoods of all other models, i.e., when

$$\forall \mathcal{M} \in \mathbb{M} \setminus \hat{\mathcal{M}} : p(\hat{\mathcal{M}} \mid \mathcal{D}) \gg p(\mathcal{M} \mid \mathcal{D}). \tag{2.56}$$

Then, a good approximation of the sum in Eq. (2.55) is given by

$$p(\mathbf{y} \mid \mathbf{x}, \mathcal{D}) \simeq p(\mathbf{y} \mid \mathbf{x}, \mathcal{D}, \hat{\mathcal{M}}). \tag{2.57}$$

Selecting the most-likely model according to Eq. (2.56) thus requires the evaluation and comparison of the posterior probabilities $p(\mathcal{M} \mid \mathcal{D})$ of all models, i.e.,

$$\hat{\mathcal{M}} = \arg \max_{\mathcal{M} \in \mathbb{M}} p(\mathcal{M} \mid \mathcal{D}). \tag{2.58}$$

By applying Bayes' rule, this term can be re-written as

$$\hat{\mathcal{M}} = \arg \max_{\mathcal{M} \in \mathbb{M}} \frac{p(\mathcal{D} \mid \mathcal{M}) p(\mathcal{M})}{p(\mathcal{D})}, \tag{2.59}$$

where the prior probability of the data $p(\mathcal{D})$ can be neglected as it is the same for all models. The prior over models, $p(\mathcal{M})$, expresses the probability with which a model $\mathcal{M}$ is expected to be the true model before having observed any data. The interesting term in Eq. (2.59) is thus the *model evidence* $p(\mathcal{D} \mid \mathcal{M})$. Given a uniform prior over the model space, the model selection problem reduces to

$$\hat{\mathcal{M}} = \arg \max_{\mathcal{M} \in \mathbb{M}} p(\mathcal{D} \mid \mathcal{M}). \tag{2.60}$$

Most models contain a parameter vector $\boldsymbol{\theta}$ over which one needs to integrate to compute the model evidence from the data likelihood, i.e.,

$$p(\mathcal{D} \mid \mathcal{M}) = \int p(\mathcal{D} \mid \mathcal{M}, \boldsymbol{\theta}) p(\boldsymbol{\theta} \mid \mathcal{M}) \, \mathrm{d}\boldsymbol{\theta}. \tag{2.61}$$

For many problems, it is reasonable to assume that the posterior distribution over the parameter vector

$$p(\boldsymbol{\theta} \mid \mathcal{M}, \mathcal{D}) \propto p(\mathcal{D} \mid \mathcal{M}, \boldsymbol{\theta}) p(\boldsymbol{\theta} \mid \mathcal{M}) \tag{2.62}$$

has a strong peak at the most-likely parameter vector $\hat{\boldsymbol{\theta}}$ so that it can be approximated with a Gaussian. By using the *Laplace approximation* (elaborated in more detail in Appendix A) we obtain

$$\log p(\mathcal{D} \mid \mathcal{M}) \simeq \log p(\mathcal{D} \mid \mathcal{M}, \hat{\boldsymbol{\theta}}) + \log p(\hat{\boldsymbol{\theta}} \mid \mathcal{M}) + \frac{k}{2} \log(2\pi) - \frac{1}{2} \log |A|, \tag{2.63}$$

where k is the number of dimensions of the parameter vector $\boldsymbol{\theta}$ and A is the Hessian of the data likelihood evaluated at $\hat{\boldsymbol{\theta}}$. In theory, the Hessian A can be estimated directly from the likelihood function. However, typically no closed form solution is available, and evaluating the Hessian is both numerically unstable and costly especially when many parameters are involved. Under a few additional assumptions (for more details, see Appendix B), the posterior of Eq. (2.63) can be approximated as

$$\log p(\mathcal{D} \mid \mathcal{M}) \simeq \log p(\mathcal{D} \mid \mathcal{M}, \hat{\boldsymbol{\theta}}) - \frac{1}{2} k \log n, \tag{2.64}$$

which is known as the *Bayesian information criterion (BIC)* (Schwarz, 1978).

The advantage of the BIC is that it is easy to compute. However, it tends to over-estimate the (effective) number of parameters and thus overly favors simple models. In practice, the BIC has been applied successfully to a large set of different model selection problems. Yet, if a more accurate estimate of the model evidence is required, one can still resort to the Laplace approximation in Eq. (2.63) or even to a full (numerical) integration of Eq. (2.61).

2.3 Graphical Models

So far, we only considered the problem of learning a model describing the relationship between two random variables. Although problems with more random variables can be treated as a single learning problem with a high-dimensional input and target space, often an internal *structure* between the random variables exists that can be exploited during model learning. In particular, if parts of the problem are conditionally independent of each other, learning them separately is much more efficient.

Probabilistic graphical models are an appealing tool to represent the dependencies between random variables in an intuitive way (Pearl, 1988; Jensen, 2001; Koller and Friedman, 2009). Graphical models encode the conditional independence structure as a graph. In general, nodes in this graph correspond to the random variables and edges (or the lack of edges) between nodes encode conditional (in-)dependencies between them.

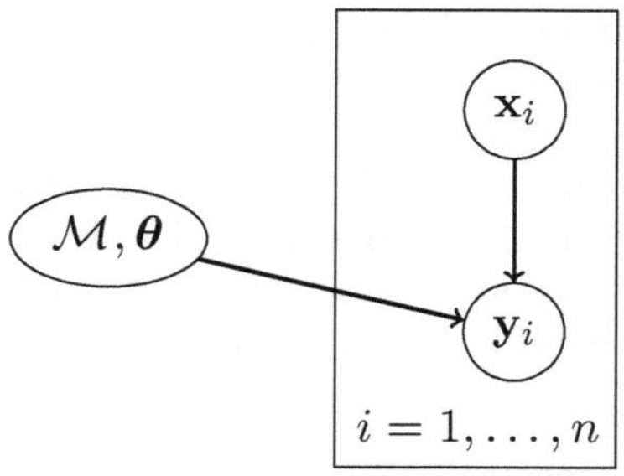

Fig. 2.6 Generic Bayesian network model underlying all regression, classification, clustering and dimensionality reduction problems discussed in this chapter.

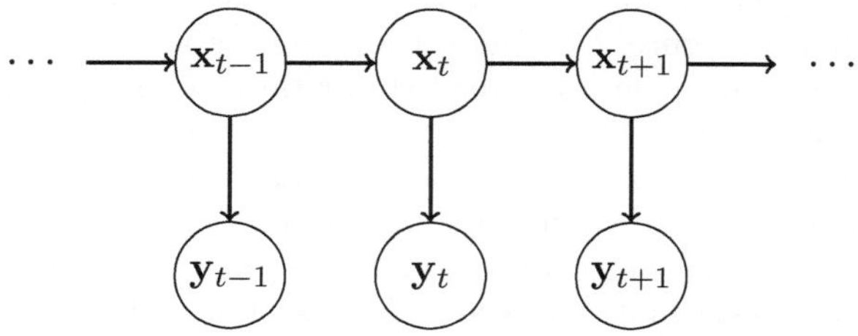

Fig. 2.7 The model underlying the Kalman filter is a dynamic Bayesian network.

Bayesian Networks

A *Bayesian network* is a graphical model that uses directed acyclic graphs to represent the dependency structure. An incoming edge indicates the conditional dependency of a child node on the node from where the edge originates (called the parent node). Nodes in a Bayesian network are conditionally independent of their ancestors given their parents, such that the joint probability distribution of all nodes is given by

$$p(\mathbf{x}_1, \ldots, \mathbf{x}_n) = \prod_{i=1,\ldots,n} p(\mathbf{x}_i \mid \mathit{parents}(\mathbf{x}_i)). \tag{2.65}$$

Here, $\mathbf{x}_1, \ldots, \mathbf{x}_n$ are the n random variables of the problem and $\mathit{parents}(\mathbf{x}_i)$ refers to the set of parents of $\mathbf{x}_i$. Therefore, specifying both the structure and the individual conditional density functions $p(\mathbf{x}_i \mid \mathit{parents}(\mathbf{x}_i))$ fully defines the joint probability distribution $p(\mathbf{x}_1, \ldots, \mathbf{x}_n)$.

In the first part of this chapter, we discussed various techniques for learning models between input and target variables. The graphical model underlying these learning problems is visualized in Figure 2.6. The training data $\mathcal{D} = \{(\mathbf{x}_i, \mathbf{y}_i)\}_{i=1}^n$ can be considered as a set of n input variables $\mathbf{x}_1, \ldots, \mathbf{x}_n$ with associated target variables $\mathbf{y}_1, \ldots, \mathbf{y}_n$. The relationship between these variables is defined by a model $\mathcal{M}$ and its parameter vector $\boldsymbol{\theta}$. Together, this induces the conditional density function $p(\mathbf{y} \mid \mathbf{x}, \mathcal{M}, \boldsymbol{\theta})$ that is shared by all observation pairs $(\mathbf{x}_i, \mathbf{y}_i)$ in the training set. Further, all training samples $(\mathbf{x}_i, \mathbf{y}_i)$ are independent of each other and identically distributed given the model $\mathcal{D}$ and its parameter vector $\boldsymbol{\theta}$. In Figure 2.6, the *plate notation* is used to indicate that the random variables $\mathbf{x}_i$ and $\mathbf{y}_i$ are copied n times, i.e., for $i = 1, \ldots, n$. As the model $\mathcal{M}$ and its parameter vector $\boldsymbol{\theta}$ are located outside the plate, they exist only once and are shared by the variables in the plate.

Dynamic Bayesian Networks

Dynamic Bayesian networks (DBN) are a special form of Bayesian networks that are well suited to represent sequences of variables. For example, the model underlying the Kalman filter is a simple DBN that represents a process as depicted in Figure 2.7. Here, the system state at time t is denoted by $\mathbf{x}_t$ and depends only on its immediate predecessor $\mathbf{x}_{t-1}$ indicated by the single, incoming arrow. The state $\mathbf{x}_t$ of the system evolves over time according to the system dynamics that are specified by the conditional density function $p(\mathbf{x}_{t+1} \mid \mathbf{x}_t)$. In each time step an observation $\mathbf{y}_t$ of the state is made according to the observation model $p(\mathbf{y}_t \mid \mathbf{x}_t)$. The Kalman filter is an efficient method to estimate the state of the system (Thrun et al., 2005).

Inference in Bayesian Networks

A typical inference problem in Bayesian networks is to infer the distribution over one or more random variables given a set of observed random variables by marginalizing over all other variables. For example, during the training phase of a regression or classification problem, this is achieved in two steps. First, the probability distribution of the model parameter $\boldsymbol{\theta}$ is inferred from the training data. Subsequently, this estimate is used for making predictions, i.e., to infer the target value of a novel input vector.

Various inference algorithms exist for Bayesian networks, including both exact and approximate methods. When all random variables are assumed to be normally distributed, often exact inference can be achieved (Koller and Friedman, 2009). Otherwise, iterative methods such as belief propagation, variational Bayes, or Markov chain Monte-Carlo sampling can be employed to approximate probability distributions over random variables in the Bayesian network. It is also possible to infer the structure of a graphical model. This includes both the connectivity between the nodes in the graph, as well as the number and dimensionality of (hidden) random variables. Depending on the problem structure, this can be implemented in a variety of ways, but the most general solution is to treat the structure as an additional (hyper-)parameter that needs to be inferred from the data. For this aim, the techniques presented in Section 2.2 can be employed. For example, alternative network structures can be ranked by their posterior probability $p(\mathcal{M} \mid \mathcal{D})$. Finally, graphical models can also be used for decision making, i.e., to control a process where outcomes are partly random and partly under the control of the decision maker.

We use graphical models in this book as a well-understood theoretical framework for modeling large probabilistic learning problems. In Chapter 3 and 4, we represent the kinematic functions of robotic manipulators and articulated objects as Bayesian networks, infer both their structure and their kinematic parameters, and use them to solve forward and inverse kinematics. In Chapter 8, we model descriptions of manipulation tasks as dynamic Bayesian networks. In this approach, we encode a task description as probability distribution over task constraints which enables a robot to reproduce a task even under different conditions.

2.4 Summary

In this chapter, we discussed the set of machine learning techniques that we use to learn flexible models for mobile manipulation robots. We introduced regression, classification, dimensionality reduction, and clustering problems and presented relevant solution techniques. Further, we showed how cross-validation and the Bayesian model posterior can be used to rank alternative models and to select the best one. Finally, we introduced Bayesian networks as a general framework to factorize large learning problems into feasible components.

Chapter 3
Body Schema Learning

Kinematic models are widely used in robotics to describe the mechanism of a robot. For example, the kinematic model of a manipulation robot is typically specified by the position of its joints, and the size and orientation of its links (Craig, 1989; Sciavicco and Siciliano, 2000). Kinematic models are usually derived analytically by a robot engineer and thus rely heavily on prior knowledge about the geometry of the robot. When such a model is applied to a real robot, its parameters have to be carefully calibrated (Gatla et al., 2007) to ensure a high accuracy, for example, using expensive calibration systems at the robot manufacturer's site. As robotic systems become more versatile and are increasingly delivered in completely reconfigurable ways, there is a growing demand for techniques to learn kinematic models automatically. Ideally, such techniques would neither require human intervention nor costly calibration equipment. This capability does not only facilitate the deployment and calibration of new robotic systems but also enables robots to autonomously adapt their models when the kinematics change, for example, as a result of hardware failures or material fatigue. Furthermore, the intelligent use of tools also requires the robot to include a tool dynamically in its kinematic model (Nabeshima et al., 2006).

The concept of kinematic models in robotics is closely related to the concept of the *body schema* in cognitive neuroscience (Stamenov, 2005; Gallagher, 2005) that refers to our internal representation of the body. Neurophysiological experiments indicate that humans as well as higher primates adapt their body schema continuously (Meltzoff and Moore, 1997), for example, when handling tools (Maravita and Iriki, 2004).

In this chapter, we develop a novel approach that allows a robot to learn its body schema using visual self-observation and exploratory actions. Our model is based on Bayesian networks that we use to represent the kinematic structure. We learn models for the individual joints of a robot using Gaussian process regression and develop an efficient algorithm to estimate the full kinematic structure of the robot. In experiments carried out in simulation and on real robots, we demonstrate that our approach enables a manipulation robot to learn its kinematic model from scratch and to maintain it over extended periods of time. Furthermore, we show that a robot using our approach can accurately predict and control the pose of its end effector even in the presence of hardware failures.

J. Sturm: *Approaches to Probabilistic Model Learning*, STAR 89, pp. 35–63.
DOI: 10.1007/978-3-642-37160-8_3

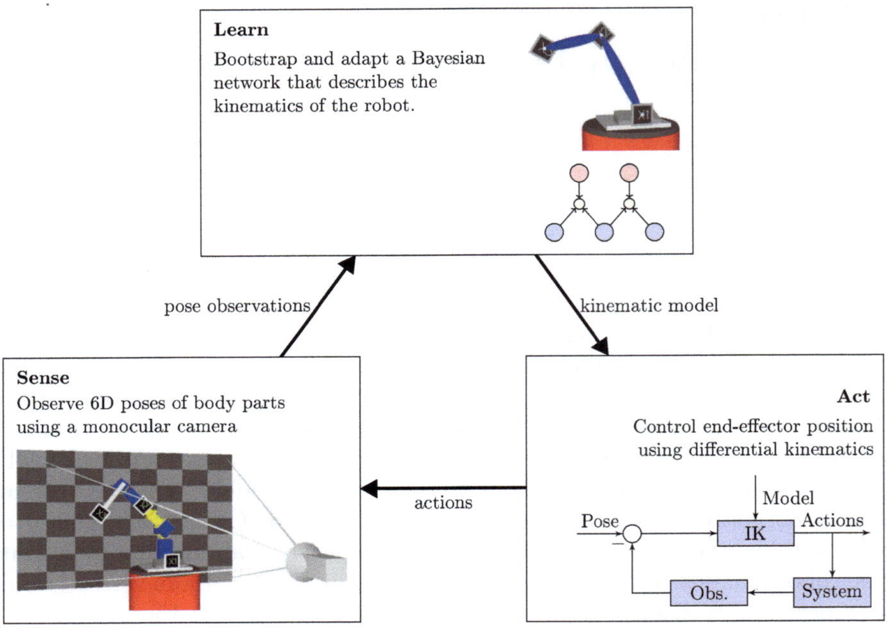

Fig. 3.1 Schematic overview of our approach to body schema learning.

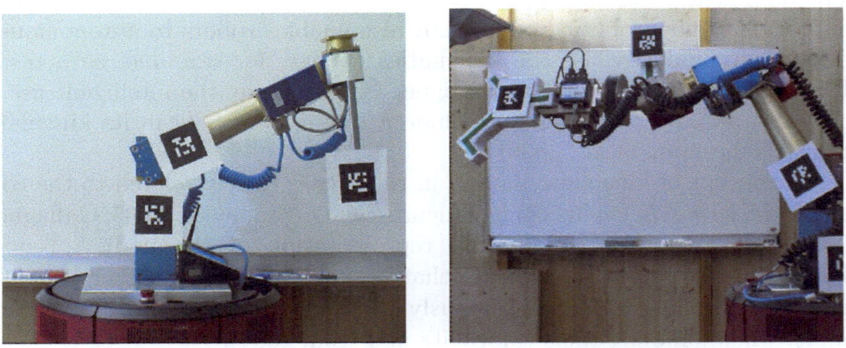

(a) 2-DOF robot (b) 7-DOF robot

Fig. 3.2 The manipulation robots used in this chapter to develop and test our approach.

Figure 3.1 illustrates the proposed approach. The robot sends random "motor babbling" commands to its joints, observes the resulting pose, and estimates the kinematic model of itself from this sequence of observations. In each iteration, the robot learns Gaussian process models for the individual joints and searches for the kinematic structure that best explains the observed motion. The robot can use the learned model to predict and control the pose of its end effector. We developed and tested our approach on several simulated and two real manipulation robots as depicted in Figure 3.2.

This chapter is structured as follows. In Section 3.1, we briefly introduce kinematic models for manipulation robots and explain how they can be represented using Bayesian networks. Subsequently in Section 3.2, we present our probabilistic framework for learning such kinematic models from visual self-observations. In Section 3.3, we extend our framework to enable a robot to localize errors in the model and efficiently replace mismatching parts. In Section 3.4, we present experimental results obtained with real and simulated manipulator arms. These experiments demonstrate that our approach is able to learn compact and accurate models and is capable of dealing robustly with noisy observations. Finally, we conclude this chapter with a discussion of related work in Section 3.5.

3.1 Kinematic Models for Manipulation Robots

The kinematic model of a manipulation robot describes the relationship between its configuration and its body posture, i.e., the relationship between the joint angles and the poses of the body parts in 3D space. Figure 3.3a shows an example of a simple manipulation robot with two degrees-of-freedom (DOF). The robot consists of two 1-DOF rotary joints $q_1, q_2 \in \mathbb{R}$, and five body parts $\mathbf{x}_1, \ldots, \mathbf{x}_5 \in SE(3)$. Here, $SE(3)$ refers to the special Euclidean group that represents all three-dimensional poses (including both position and orientation). Internally, we chose to represent these 3D poses as homogeneous $\mathbb{R}^{4\times 4}$ matrices, which can be concatenated and inverted, but other representations such as translation vectors and quaternions or twists are also frequently used (Ma et al., 2003). The first two body parts are connected rigidly. This means that the geometric transformation $\boldsymbol{\Delta}_{12} \in SE(3)$ from the trunk $\mathbf{x}_1$ to the shoulder $\mathbf{x}_2$ is independent of the configuration of the joints. The shoulder $\mathbf{x}_2$ and the upper arm $\mathbf{x}_3$ are connected by the shoulder joint q_1, and thus their geometric transformation $\boldsymbol{\Delta}_{23}(q_1) : \mathbb{R} \mapsto SE(3)$ depends on the joint angle of q_1. The same holds for the following parts, as the joint angle of the elbow joint q_2 has direct influence on the geometrical transformation $\boldsymbol{\Delta}_{34}(q_2) : \mathbb{R} \mapsto SE(3)$ between the upper arm $\mathbf{x}_3$ and the lower arm $\mathbf{x}_4$. The gripper $\mathbf{x}_5$ is attached rigidly to the lower arm $\mathbf{x}_4$, such that $\boldsymbol{\Delta}_{45} \in SE(3)$ is a fixed transformation. The kinematic function of this manipulator can thus be constructed by the concatenation of these individual transforms, i.e.,

$$f(q_1, q_2) := \boldsymbol{\Delta}_{12} \circ \boldsymbol{\Delta}_{23}(q_1) \circ \boldsymbol{\Delta}_{34}(q_2) \circ \boldsymbol{\Delta}_{45}. \tag{3.1}$$

The kinematic function $f(q_1, q_2)$ describes the full geometrical transformation from the coordinate frame of the trunk to the coordinate frame of the gripper.

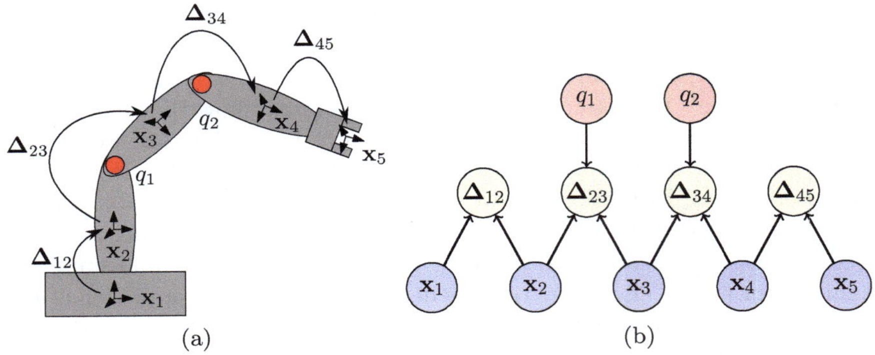

Fig. 3.3 (a) Simple 2-DOF manipulator consisting of 5 body parts. (b) The kinematic model of this robot represented as a Bayesian network.

Denavit-Hartenberg Convention

In engineering, the kinematic function of a serial manipulation robot is often defined as a parameterized model $f_{\mathcal{M},\boldsymbol{\theta}}$ where the parameters $\boldsymbol{\theta}$ of the individual transformations are specified using the *Denavit-Hartenberg (DH)* convention (Sciavicco and Siciliano, 2000). In this convention, each link is specified with four parameters $d_i, \theta_i, r_i, \alpha_i$ and constructed as a screw displacement (rotation and translation) around/along the Z axis and a second screw displacement along/around the X axis:

$$\boldsymbol{\Delta}_{i(i+1)} = \text{Rot}_Z(\theta_i)\text{Trans}_Z(d_i)\text{Trans}_X(r_i)\text{Rot}_X(\alpha_i). \tag{3.2}$$

For a rotary joint, θ_i varies with the joint angle q_i while for a prismatic joint, d_i varies with the joint position q_i. All other parameters remain fixed when the manipulator moves. In sum, a robot consisting of n joints can thus be described by a $(4n)$-dimensional parameter vector $\boldsymbol{\theta} \in \mathbb{R}^{4n}$. Additional parameters can be introduced as needed, for example for sensing and transmission elements such as encoder offsets or gear ratios.

Kinematic Calibration

While an initial guess of the kinematic parameters is in general available from the construction plan, accurate positioning requires fine-tuning of the parameters by means of kinematic calibration or, more generally, system identification (Hollerbach et al., 2008). To this aim, typically a large set of measurements is obtained while the manipulator approaches different configurations. Depending on the calibration setup, the measurements can be full pose observations $\mathbf{z}_{ij} \in SE(3)$ of individual links $\boldsymbol{\Delta}_{ij}$ or the whole arm (Pradeep et al., 2010). Alternatively, only parts of the pose can be observed, for example, 3D position, 2D bearing, or 1D distance using a laser tracker. Furthermore, the end-effector of the manipulator can be physically constrained, for example, by making contact with a fixed point in the environment. Hollerbach and Wampler (1996) presented a widely adopted approach termed the calibration index that treats all pose measurements and physical constraints in a unified framework. The general idea is to formulate all measurements and constraints as a system of (nonlinear) equations

$$g(\boldsymbol{\theta}) = (g_1^T(\boldsymbol{\theta}), \ldots, g_P^T(\boldsymbol{\theta}))^T = \mathbf{0}, \tag{3.3}$$

where each $g_i(\boldsymbol{\theta})$ corresponds to one measurement equation. For example, for a 3D position observations, these functions could be defined as

$$g_i(\boldsymbol{\theta}) = \mathbf{z}_i - \text{trans}(f_{\mathcal{M},\boldsymbol{\theta}}(\mathbf{q}_i)) \tag{3.4}$$

where $\mathbf{z}_i$ is the observed 3D position corresponding to configuration $\mathbf{q}_i$, and trans($\cdot$) extracts the translational components of the pose predicted by the kinematic model $f_{\mathcal{M},\boldsymbol{\theta}}$. This system of equations can be efficiently solved using the nonlinear least squares minimization techniques presented in Chapter 2 such as the Levenberg-Marquardt algorithm.

Inverse Kinematics

For many robotic applications, it is necessary to compute the robot configuration $\mathbf{q}$ to reach a given target position in the workspace. This requires the inversion of f, which is also called the inverse kinematic function. As the algebraic inversion is only possible for simple manipulators, a solution to the inverse kinematic problem is in practice often computed using an iterative numerical method such as the Jacobian transpose, pseudo-inverse or damped-least squares method (Buss and Kim, 2005; Sciavicco and Siciliano, 2000).

Bayesian Networks

A fundamental insight in our work is that the kinematic model of a manipulation robot (and also those of articulated objects as we will see in Chapter 4) can be represented in form of Bayesian networks. Consider the example given in Figure 3.3b: the configuration variables q_1 and q_2, the poses of the body parts $\mathbf{x}_1, \ldots, \mathbf{x}_5$, and the relative transformations $\boldsymbol{\Delta}_{12}, \ldots, \boldsymbol{\Delta}_{45}$ of our example robot appear as nodes in the Bayesian network. Further, the topology of the network encodes the kinematic structure: the relative transformation $\boldsymbol{\Delta}_{12}$ relates the first two body parts $\mathbf{x}_1$ and $\mathbf{x}_2$, while the second relative transformation $\boldsymbol{\Delta}_{23}$ depends additionally on the configuration q_1 of the first joint.

We can now use standard inference techniques for Bayesian networks to predict the pose of the end effector (given $q_1, \ldots, q_m$ and $\mathbf{x}_1$, infer $\mathbf{x}_n$) or to control the pose of the end effector (given $\mathbf{x}_1$ and $\mathbf{x}_n$, infer $q_1, \ldots, q_m$). Both problems can be solved by marginalizing over all other variables in the network: solving forward kinematics corresponds to a marginalization over all intermediate body parts. As we will elaborate in the next section, this marginalization can be solved efficiently and in closed form when we assume that all variables in the Bayesian network are normally distributed.

3.2 A Bayesian Framework for Body Schema Learning

One of the central ideas in our work is to use Bayesian networks for representing kinematic models. We define the robotic body schema as the joint probability distribution over joint actions $\mathbf{q} = (q_1, \ldots, q_m)$, true poses $\mathbf{x} = (\mathbf{x}_1, \ldots, \mathbf{x}_n)$, and pose observations $\mathbf{y} = (\mathbf{y}_1, \ldots, \mathbf{y}_n)$ of a manipulation robot. The individual $q_i \in \mathbb{R}$ are real-valued variables corresponding to the latest configuration request sent to the i-th joint of the robot. The $\mathbf{x}_i \in SE(3)$ encode the true poses of the body parts with respect to a reference coordinate frame. The $\mathbf{y}_i \in SE(3)$ are the robot's pose observations of its body parts that are generally noisy and potentially missing. We denote a sequence of t action-pose observations as $\mathcal{D} = \langle(\mathbf{q}^1, \mathbf{y}^1), (\mathbf{q}^2, \mathbf{y}^2), \ldots, (\mathbf{q}^t, \mathbf{y}^t)\rangle$. Formally, we seek to learn the probability distribution

$$p(\mathbf{x}_1, \ldots, \mathbf{x}_n, \mathbf{y}_1, \ldots, \mathbf{y}_n \mid q_1, \ldots, q_m), \tag{3.5}$$

which in this form is intractable for all but the simplest scenarios. Therefore, we assume that each observation variable $\mathbf{y}_i$ is independent from all other variables given the true pose $\mathbf{x}_i$ of the corresponding body part and that they can thus be fully characterized by an observation model $p(\mathbf{y}_i \mid \mathbf{x}_i)$. Furthermore, if the kinematic structure of the robot was known, a large number of pair-wise independencies between action signals and body parts could be assumed, which in turn would lead to the much simpler, factorized model

$$p(\mathbf{x}_1, \ldots, \mathbf{x}_n \mid q_1, \ldots, q_m) = \prod_i p(\mathbf{x}_i \mid \text{parents}(\mathbf{x}_i)). \tag{3.6}$$

Here, parents($\mathbf{x}_i$) refers to the parent nodes of $\mathbf{x}_i$ in the Bayesian network and comprises only those body parts and action signals on which $\mathbf{x}_i$ directly depends on. Note that the actions are given and, thus, do not depend on other variables in this model. We now make the factorized structure of the problem explicit by introducing hidden variables $\mathbf{\Delta}_{ij} := \mathbf{x}_i^{-1}\mathbf{x}_j$ corresponding to the relative geometric transformation between all pairs $(\mathbf{x}_i, \mathbf{x}_j)$ of body parts. Further, we denote with $\mathbf{z}_{ij} := \mathbf{y}_i^{-1}\mathbf{y}_j$ the relative geometric transformation relating the observations $\mathbf{y}_i$ and $\mathbf{y}_j$ that correspond to $\mathbf{x}_i$ and $\mathbf{x}_j$. Using this, we define as a *local model* the subgraph of our network that describes the geometric relationship between any two body parts $\mathbf{x}_i$ and $\mathbf{x}_j$ given the relevant part of the action signal, if all other body parts are ignored. Figure 3.4 shows a prototypical local model. Here, we denote with $\mathcal{Q}_{ij}$ the set of action signals that have a direct influence on $\mathbf{\Delta}_{ij}$. Any set of $(n-1)$ local models which forms a spanning tree over all n body parts defines a model for the whole kinematic structure and is a solution to Eq. (3.6).

Note that our approach does not depend on a proprioceptive sensor telling the robot in which configuration a particular joint is after executing an action q_i. At first sight, it seems that with proprioception one could learn the kinematic function passively from visual and proprioceptive observations only. While this is true, one would lack the mapping from motor commands to motor encoders such that the learned model would not suffice for manipulator control. One would either need to assume that motors and proprioceptive sensors are calibrated precisely, or one would need to additionally learn the mapping from actions to joint encoder values for each joint. In contrast to this, we learn a combined model that directly maps from motor commands to body pose observations. In this way, our approach closes the action-perception-loop, as visualized in Figure 3.1, and it obviates the need for the explicit calibration of the motor encoders. For the sake of completeness, it should be noted that our approach can also be used to learn the kinematic model based on proprioception, by replacing the motion requests by the observed joint configurations.

In the following, we explain how to learn local models from data and how to find the spanning tree built from these local models that best explains the whole robot. We consider the single best solution only and do not perform model averaging over possible alternative structures. Note that in theory, it would be straight-forward to keep multiple structure hypotheses and to average over them for prediction using Bayes' rule. Control under structure uncertainty, however, is a slightly more difficult problem. One would have to consider all possible structures and assess the individual risks and gains for alternative actions. Then, one would select the action that maximize the overall gain while keeping all possible risks low. In practice, we found that considering the most-likely structure only is sufficient for most of the relevant tasks. Our approach is conservative in this respect since it requires a certain minimal accuracy from all parts of the body schema before the model is considered complete.

3.2.1 Local Models

The local kinematic models are the central concept in our body schema framework. A *local model* $\mathcal{M}$ (see Figure 3.4) describes the geometric relationship between two body parts i and j given a set of action signals $\mathcal{Q}_{ij}$. We propose to learn this relationship from data samples acquired while requesting random joint configurations and observing their effects on the robot's pose. As the learning framework for solving this supervised regression problem, we apply Gaussian process models for regression (Rasmussen and Williams, 2006). The observations $\mathbf{y}_i$ of part locations $\mathbf{x}_i$ are obtained by tracking visual markers in 3D space including their position and orientation (Fiala, 2005). These markers are also depicted in Figure 3.2. Note that the observations $\mathbf{y}_i$'s are inherently noisy and that missing observations are common, for example, in consequence of (self-)occlusion. Formally, the task is to learn the local transformations $\mathbf{\Delta}_{ij}$, each linking two body parts $\mathbf{x}_i$ and $\mathbf{x}_j$. Considering Figure 3.4, a straight-forward approach would be to infer the true poses $\mathbf{x}_i$ and $\mathbf{x}_j$ from the noisy observations $\mathbf{y}_i$ and $\mathbf{y}_j$, by assuming Gaussian white noise on the observations, i.e.,

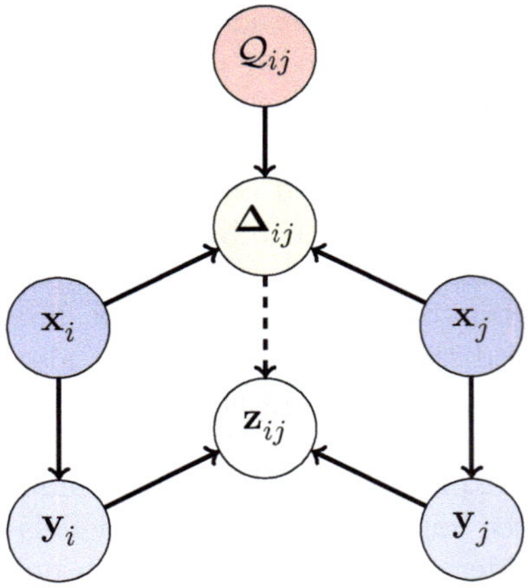

Fig. 3.4 Template of a local model that defines the kinematics between two related body parts.

$$\mathbf{y}_i \sim \mathcal{N}(\mathbf{x}_i, \Sigma_{\mathbf{y}}). \tag{3.7}$$

Then, one would need to integrate over the latent true poses $\mathbf{x}_i$ and $\mathbf{x}_j$ in order to reason about $\boldsymbol{\Delta}_{ij}$.

However, since the *absolute* positions $\mathbf{x}_i$ are irrelevant for describing the *relative* transformations, we take a slightly different approach by focusing directly on the transformations $\mathbf{z}_{ij}$ between observations $\mathbf{y}_i$ and $\mathbf{y}_j$. Note that these virtual measurements $\mathbf{z}_{ij}$ are noisy observations of the true transformation Δ_{ij} as a result of Eq. (3.7), i.e., we obtain

$$\mathbf{z}_{ij} \sim \mathcal{N}(\boldsymbol{\Delta}_{ij}, \Sigma_{\mathbf{z}}). \tag{3.8}$$

With this, we can directly learn the relationships of actions $\mathcal{Q}_{ij}$ to relative transformations $p(\mathbf{z}_{ij} \mid \mathcal{Q}_{ij})$. The problem of learning a single local model now has the form of the noisy regression problem

$$\mathbf{z}_{ij} = f_{\mathcal{M}}(\mathcal{Q}_{ij}) + \boldsymbol{\epsilon} \tag{3.9}$$

that is, the regression function

$$\begin{aligned} f_{\mathcal{M}}: \quad \mathbb{R}^{|\mathcal{Q}_{ij}|} &\rightarrow \mathbb{R}^{16}, \\ \mathcal{Q}_{ij} &\mapsto \Delta_{ij} \end{aligned} \tag{3.10}$$

has to be learned from a sequence of noisy observations $\mathbf{z}_{ij}$.

For simplicity, we consider the over-parametrized transformation matrices in the following with $d = 12$ independent components and keep the remaining 4 elements of the homogeneous matrices fixed to (0 0 0 1). Subsequently, we learn the functional mapping for each of the 12 components separately. Due to this simplification, we cannot guarantee that all predictions correspond

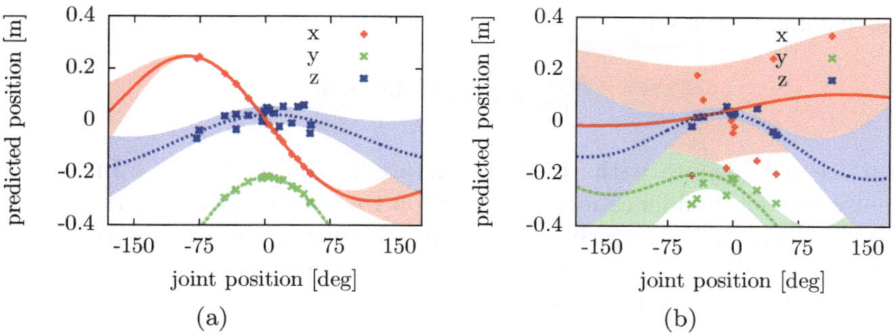

Fig. 3.5 Two local models learned from real data. (a) Example of an accurate local model. (b) Another local model that is less likely to be selected. The shaded areas represent the uncertainty of the learned Gaussian process.

to valid, homogeneous transformation matrices. In practice, however, they lie close to valid transformations such that a normalization step resolves the problem. In particular, we ortho-normalize the rotational part of the homogeneous matrix using singular value decomposition. For solving the regression problem as stated in Eq. (3.10), we learn a Gaussian process model (Rasmussen and Williams, 2006) for the transformation functions $f_{\mathcal{M}}$ for all local models $\mathcal{M}$ and choose the squared exponential covariance function to parametrize the process.

An example of this is given in Figure 3.5. The red, green and blue curves show the translational x-, y-, and z- components of two different local models, respectively. The depicted models were learned from real data using Gaussian process regression. In the situation shown in Figure 3.5a, the action (x-axis) physically corresponds to the transformation being measured (y-axis). Thus, the data set is self-consistent and accurate functions with low noise levels can be learned. The higher noise level for the z-component is due to larger measurement error in this direction (i.e., the camera's line of vision). In the situation depicted in Figure 3.5b, a local model has been learned for variables that do not have a direct physical relationship. As a result, the model predicts the observations with a high uncertainty and thus does not explain the data well. Such a local model is likely to be discarded during the search for the full body model.

3.2.2 Learning a Factorized Full Body Model

We seek to find a factorized model of the kinematic model that best explains the observed data. Our aim is to learn and evaluate this model efficiently, i.e., we aim to minimize the number of local models that need to be learned.

We implement this by discarding all local models that are overly inconsistent with the observed data. We define a local model $\mathcal{M}$ to be valid given a set of observations $\mathcal{D}$, if and only if the sample observation log-likelihood is above some threshold η, i.e.,

$$\frac{1}{|\mathcal{D}|} \log p(\mathcal{D} \mid \mathcal{M}) > \eta \tag{3.11}$$

that we will denote with the Boolean predicate $valid_{\mathcal{M}}(\mathcal{D})$. In practice, we use the 3σ confidence interval based on the sensor noise as a threshold to reject models that are overly inconsistent with the observations. We compute the data likelihood of a set of observations $\mathcal{D}$ as the product of the likelihoods of the individual observations, i.e.,

$$p(\mathcal{D} \mid \mathcal{M}) := \prod_{(\mathbf{z}_{ij}, \mathcal{Q}_{ij}) \in \mathcal{D}} p(\mathbf{z}_{ij} \mid \mathcal{Q}_{ij}, \mathcal{M}). \tag{3.12}$$

According to our observation model from Eq. (3.8), we assume Gaussian noise in the observations $\mathbf{z}_{ij}$ with covariance $\Sigma_{\mathbf{y}}$ with respect to the expected pose $\hat{\Delta}_{ij} := \mathbb{E}[\Delta_{ij} \mid \mathcal{Q}_{ij}, \mathcal{M}]$ as predicted from the Gaussian process model, resulting in

$$p(\mathbf{z}_{ij} \mid \mathcal{Q}_{ij}, \mathcal{M}) := \frac{1}{\sqrt{(2\pi)^6 \, |\Sigma_{\mathbf{y}}|}} \exp\left(-\frac{1}{2} (\mathbf{z}_{ij} - \hat{\Delta}_{ij})^T \Sigma_{\mathbf{y}} (\mathbf{z}_{ij} - \hat{\Delta}_{ij}) \right). \tag{3.13}$$

To compare models with different data likelihoods and complexities, we define a *model quality measure* as

$$q(\mathcal{M}) := \underbrace{\log p(\mathcal{D} \mid \mathcal{M})}_{\text{accuracy}} - \underbrace{k \log(\eta \, |\mathcal{D}|)}_{\text{complexity}} \tag{3.14}$$

where $k \in \mathbb{N}$ denotes the dimensionality of the model $\mathcal{M}$, i.e., the number $|\mathcal{Q}_{ij}|$ of action signals that the model depends on. This measure is proportional to both the model accuracy and to a penalty term for model complexity. Note that this quality measure is similar to the Bayesian information criterion (BIC) as introduced in Chapter 2. The key difference of our quality measure is that it contains the likelihood threshold as an additional factor in the complexity penalty. This formulation provides us two important properties that we can exploit to specify an efficient search strategy for the kinematic structure. These two properties are:

- Given two models of the same complexity but different data likelihoods, the quality measure favors the model with the better data fit.
- Given two valid models with different complexity, the quality measure favors the model with the lower complexity.

The first property follows directly from the definition of the quality measure. The second property results from the definition of valid models in Eq. (3.11) in combination with the threshold as a factor in the model quality measure.

If $k_1 < k_2$ and both models are valid, i.e., both $\log p(\mathcal{D} \mid \mathcal{M}_1) > \eta|\mathcal{D}|$ and $\log p(\mathcal{D} \mid \mathcal{M}_2) > \eta|\mathcal{D}|$, we can show that $q(\mathcal{M}_1) > q(\mathcal{M}_2)$ as follows:

$$\begin{aligned}
q(\mathcal{M}_1) - q(\mathcal{M}_2) &= \log p(\mathcal{D} \mid \mathcal{M}_1) - k_1 \log(\eta\,|\mathcal{D}|) - [\log p(\mathcal{D} \mid \mathcal{M}_2) - k_2 \log(\eta\,|\mathcal{D}|)] \\
&> \log \eta\,|\mathcal{D}| - k_1 \log(\eta\,|\mathcal{D}|) - [\log p(\mathcal{D} \mid \mathcal{M}_2) - k_2 \log(\eta\,|\mathcal{D}|)] \\
&\geq \log \eta\,|\mathcal{D}| - k_1 \log(\eta\,|\mathcal{D}|) - [\log 1 - k_2 \log(\eta\,|\mathcal{D}|)] \\
&\geq \log \eta\,|\mathcal{D}| - k_1 \log(\eta\,|\mathcal{D}|) - [\log 1 - (k_1 + 1) \log(\eta\,|\mathcal{D}|)] \\
&= \log \eta\,|\mathcal{D}| - k_1 \log(\eta\,|\mathcal{D}|) - 0 + k_1 \log(\eta\,|\mathcal{D}|) + \log(\eta\,|\mathcal{D}|) = 0.
\end{aligned}$$

Finding the Network Topology

If no prior knowledge about the body structure of the robot exists, we initialize a fully connected kinematic model containing a total of $\sum_{k=0}^{m} \binom{n}{2}\binom{m}{k}$ local models (linking m action signals to n relative transformations). Given a set of observations, the robot first eliminates those local models that are highly inconsistent with the data by evaluating $valid_{\mathcal{M}}(\mathcal{D})$ as described above. The remaining set of valid models is typically still large. Certain ambiguities will, for instance, remain even after infinitely many training samples. If, for example, $p(\mathbf{z}_{12} \mid q_1, \mathcal{M}_1)$ has been determined to be a valid local model, then $p(\mathbf{z}_{12} \mid q_1, q_2, \mathcal{M}_2)$ will also be. Although these alternative models might not be distinguishable regarding their data likelihood $p(\mathcal{D} \mid \mathcal{M})$, they differ significantly in their complexities k and therefore in their model quality $q(\mathcal{M})$.

To find the best topology on a global level, we aim to select the minimal subset $\hat{\mathbb{M}} \subset \mathbb{M}_{\text{valid}}$ from the superset of all valid local models $\mathbb{M}_{\text{valid}} = \{\mathcal{M}_1, \ldots \mathcal{M}_N\}$ that covers all body parts and simultaneously maximizes the overall model fit, i.e.,

$$\hat{\mathbb{M}} := \arg\max_{\mathbb{M}} \sum_{\mathcal{M} \in \mathbb{M}} q(\mathcal{M}). \tag{3.15}$$

This subset can be found efficiently by computing the minimal spanning tree of $\mathbb{M}_{\text{valid}}$ taking the negative model quality measure of the individual local models as cost function. For our purposes, the spanning tree needs to cover all body parts but not necessarily all action variables, since some of them might not have an influence on the robot.

To connect all n body poses in the Bayesian network, exactly $|\hat{\mathbb{M}}| = (n - 1)$ local models need to be selected. This yields $\binom{|\mathbb{M}_{\text{valid}}|}{|\hat{\mathbb{M}}|}$ possible network structures to be considered. In the typical case, where the robot is composed of $n-1$ arbitrarily connected 1-DOF joints, this number reduces to the order of $O(n^3)$. Regarding the scalability to higher degrees of freedom and longer kinematic chains, the growth of the search space is of less practical importance than other factors such as the *observability* of local transformations (from a given camera view point).

We illustrate our approach with an example. Figure 3.6 shows a simulated robot consisting of two body parts $\mathbf{x}_1$ and $\mathbf{x}_2$ linked by a 2-DOF spherical

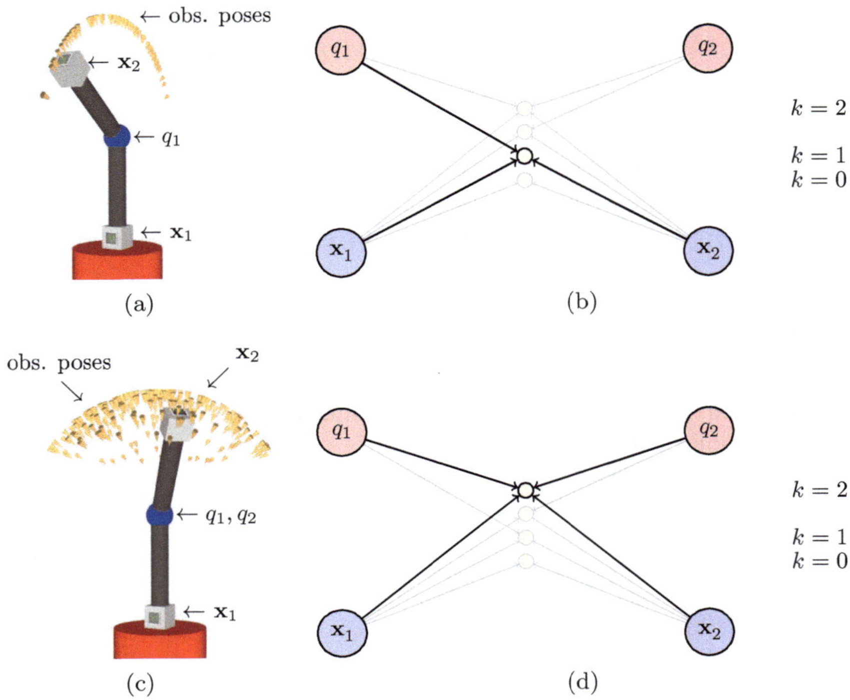

Fig. 3.6 Example of a 2-DOF robot composed of two body parts and a single spherical joint. (a)+(b) Result after actuating only the first DOF. (c)+(d) Result after actuating both DOF.

joint with two action signals $\mathbf{q} = (q_1\ q_2)^T$. To learn its kinematic model, the robot repeatedly samples random actions $\mathbf{q}$ and sends these to its joint. After the motion comes to rest, the robot observes the resulting pose of its body parts and adds the action-pose pair to the sequence of training data. Given these pose observations, it learns four local models relating its two body parts, for all possible dependencies on the two action signals: the first model is independent of any action signal ($k = 0$), the second and third model depend on one action signal ($k = 1$) q_1 and q_2, respectively, and the fourth model on both action signals ($k = 2$). Initially, we let the robot only actuate the first DOF q_1 and keep $q_2 = 0$ fixed. Correspondingly, the robot moves its end effector on a circular arc, as visualized by the yellow cones above the robot in Figure 3.6a. From this data, the robot trains all four local models. After learning, both models $\mathcal{M}_2$ and $\mathcal{M}_4$ are evaluated to be valid, i.e., have $\log p(\mathcal{D} \mid \mathcal{M}) > \eta|\mathcal{D}|$. With respect to our quality measure, however, $\mathcal{M}_4$ has a much higher complexity penalty as $k_2 = 1$ and $k_4 = 2$, and correspondingly, $\mathcal{M}_2$ is selected. The resulting kinematic structure is visualized by the bold arrows in Figure 3.6b. This situation looks different when the robot actuates both DOFs simultaneously. The resulting area covered by the end effector

Algorithm 1. Estimation of the kinematic structure

Input: training data $\mathcal{D}$
Output: kinematic structure $\hat{\mathbb{M}}$
1 **for** $k \in \{0, 1, \ldots, m\}$ **do**
2 Let $\mathbb{M}_k := \{\mathcal{M} \mid \log p(\mathcal{D} \mid \mathcal{M}) > \eta|\mathcal{D}| \wedge |\mathcal{Q}| = k\}$ be the set of all valid models of complexity k;
3 Let $\mathbb{M}_{1:k} := \bigcup_{i=1}^{k} \mathbb{M}_i$ be the set of all valid models found so far;
4 **if** *a spanning tree of* $\mathbf{x}_1, \ldots, \mathbf{x}_n$ *exists in* $\mathbb{M}_{1:k}$ **then**
5 Compute the minimum spanning tree $\hat{\mathbb{M}}$ from $\mathbb{M}_{1:k}$, for example, using Prim's or Kruskal's algorithm;
6 Return $\hat{\mathbb{M}}$ as the optimal kinematic structure;
7 **end**
8 **end**

then corresponds to a hemisphere, as visualized in Figure 3.6c. Again, the robot trains all possible local models, but now finds that only $\mathcal{M}_4$ is valid (see Figure 3.6d). As $\mathcal{M}_2$ does not depend on the second DOF, its data likelihood is far below the acceptance threshold η and thus gets rejected. These two examples demonstrate that our quality measure favors simple models over more complex ones, but also selects more complex models if necessary.

Note that for implementing this structure search efficiently, typically not every of the $\sum_{k=0}^{m} \binom{n}{2}\binom{m}{k}$ possible local models needs to be evaluated. By the choice of the quality measure in Eq. (3.14), a valid model with a lower complexity will always have a higher quality than any other valid model with a larger complexity. This follows from the threshold on valid models which serves as a lower bound on the model quality: all models with data likelihoods below this threshold are invalid and thus discarded. As a consequence, an efficient algorithm can be devised to minimize the number of models to be evaluated. It is sufficient to evaluate only the first k complexity layers of local models until a minimal spanning tree is found for the first time. This spanning tree then corresponds to the global maximum of the overall model quality. The resulting algorithm is given in Algorithm 1. Important for the efficiency is that only the minimal set of local models actually gets trained and evaluated (line 2–3) and that the algorithm stops training more models after the first spanning tree has been found (line 4–6).

We illustrate the effect of this property in Figure 3.7. In this experiment, we consider a manipulator consisting of five body parts and four action signals. The yellow nodes correspond to all theoretically possible local models. The local models depicted in this figure are sorted corresponding to their complexity, i.e., the bottommost row corresponds to local models representing rigid transforms ($k = 0$), the four next rows correspond to local models that depend only on a single action signal ($k = 1$), the next six rows to models that depend on two action signals simultaneously ($k = 2$), and so on. After the robot has evaluated the first two complexity layers ($k = 0$ and $k = 1$),

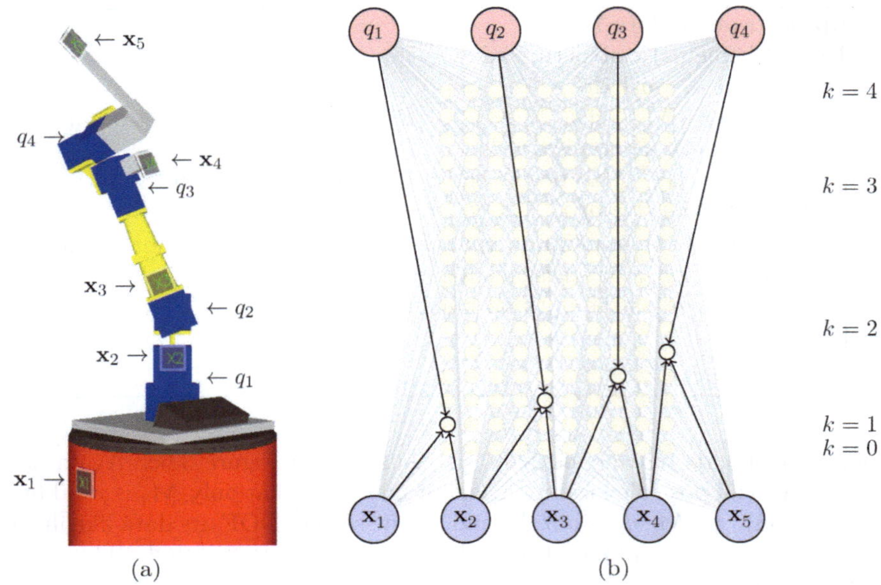

Fig. 3.7 (a) Example of a 4-DOF serial chain manipulator consisting of five body parts. (b) Recovered kinematic model.

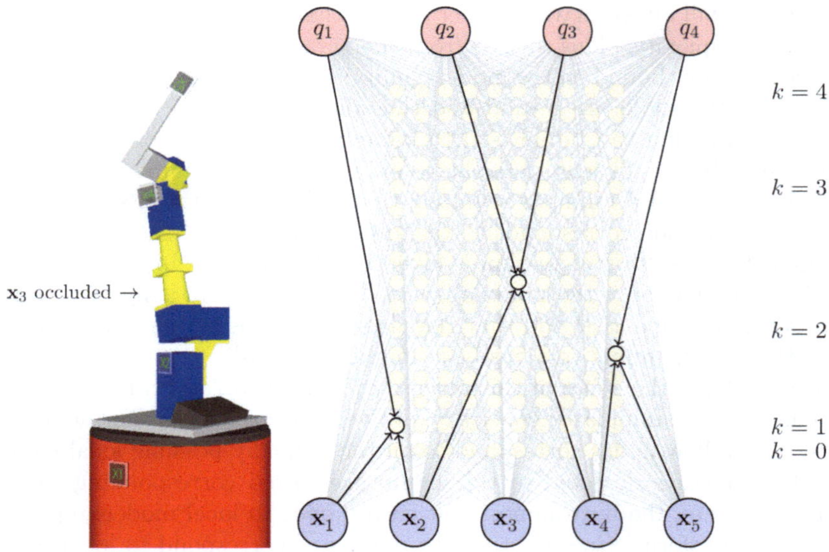

Fig. 3.8 Same robot as in Figure 3.7, but $\mathbf{x}_3$ was occluded and thus never observed. As a result, a joint model from $\mathbf{x}_2$ to $\mathbf{x}_4$ depending both on q_2 and q_3 is selected.

it detects that the set of valid models contains a spanning tree, and thus the evaluation of all remaining local models with $k \geq 2$ can be skipped. The best kinematic model corresponds to the minimum spanning tree between all body parts and the local models and is visualized by the bold edges in the figure. This experiment illustrates that the proposed quality measure contributes to the efficiency of our approach, as only the first two layers of local models need to be evaluated to find the optimal kinematic model.

In a second experiment, we occluded the visual marker corresponding to the third body part of the same robot. Figure 3.8 shows the resulting Bayesian network. As $\mathbf{x}_3$ was never observed, no local model relating the other body parts to $\mathbf{x}_3$ could be trained. Therefore, after evaluating the local models with complexities $k = 0$ and $k = 1$, no spanning tree exists, as no valid connection between $\mathbf{x}_2$ and $\mathbf{x}_4$ can be established. Only after evaluating additionally all local models that simultaneously depend on two action signals, the robot finds a local model between $\mathbf{x}_2$ and $\mathbf{x}_4$ depending both on $\mathbf{q}_2$ and $\mathbf{q}_3$. This experiment demonstrates that our approach also works when only parts of the system are observable. However, learning local models with high-dimensional inputs is a more complex learning problem and usually requires more training samples before the same prediction accuracy is achieved.

3.2.3 Pose Prediction and End-Effector Pose Control

Having discussed the learning of local models and the selection of the network structure, we now show how the resulting model can be used to predict the pose of the robot for a given action (forward kinematics) and how to infer a suitable action that moves the manipulator to a given pose (inverse kinematics).

The *kinematic forward model* can be constructed directly from the local models contained in $\mathbb{M}$, since these form a tree over all body part variables $\mathbf{x}_i$. We can write

$$p(\mathbf{x}_1, \ldots, \mathbf{x}_n \mid q_1, \ldots, q_m) = \prod_i p(\mathbf{x}_i \mid \text{parents}(\mathbf{x}_i)) \tag{3.16}$$

$$= p(\mathbf{x}_{\text{root}}) \prod_{\mathcal{M}_{ij} \in \mathbb{M}} p(\Delta_{ij} \mid \mathcal{Q}_{ij}, \mathcal{M}_{ij}) \tag{3.17}$$

$$= p(\mathbf{x}_{\text{root}}) \prod_{\mathcal{M}_{ij} \in \mathbb{M}} p(\mathbf{x}_i^{-1} \mathbf{x}_j \mid \mathcal{Q}_{ij}, \mathcal{M}_{ij}) \ , \tag{3.18}$$

where $\mathbf{x}_{\text{root}}$ is the position of the robot trunk, which serves as the reference frame for all other body parts. We use $\mathcal{M}_{ij}$ to denote the local model of $\mathbb{M}$ which describes the transformation between $\mathbf{x}_i$ and $\mathbf{x}_j$. From $p(\mathbf{x}_1, \ldots, \mathbf{x}_n | q_1, \ldots, q_m)$ in the factorized form, we can now approximate the maximum likelihood estimate of the resulting body posture given an action $\mathbf{q}$ by concatenating the geometric transformations of the individual geometric transformations. We define the kinematic function by finding the maximum of the probability distribution

$$f(\mathbf{q}) := \max_{\mathbf{x}_{\text{ee}}} p(\mathbf{x}_n \mid q_1, \ldots, q_m, \mathbf{x}_{\text{root}}), \tag{3.19}$$

where $\mathbf{x}_{\text{ee}}$ denotes the body part corresponding to the end effector (i.e., the body part to be controlled). As all local models evaluated for a particular action $\mathbf{q}$ provide a Gaussian distribution in pose space, the marginal over the pose of the end effector can efficiently be computed as the concatenation of the marginals of the individual local models. In particular, we are interested in the maximum likelihood estimate for the end effector which we can compute efficiently by concatenation, i.e.,

$$f(\mathbf{q}) := f_{\mathcal{M}_{12}}(\mathcal{Q}_{12}) f_{\mathcal{M}_{23}}(\mathcal{Q}_{23}) \cdots f_{\mathcal{M}_{(n-1)n}}(\mathcal{Q}_{(n-1)n}). \tag{3.20}$$

Here, $f_{\mathcal{M}_{ij}}(\mathcal{Q}_{ij})$ refers to transformation predicted by the local model $\mathcal{M}_{ij}$ and evaluated for relevant part of the action signal $\mathcal{Q}_{ij}$. Note that also the covariances of the pose estimate can be computed efficiently by approximating the result of the multiplication of two Gaussians with a Gaussian. As each regression function $f_{\mathcal{M}_{ij}}$ corresponds to a Gaussian process, also the expected variance is known and can be propagated efficiently, similar to Eq. (3.20), through the Bayesian network. We may refer the interested reader to Ware and Lad (2003) on this topic. In practice, however, we found that estimating the variance directly from the training data is more reliable, as it provides us with a global estimate of the uncertainty instead of a summation over local uncertainties.

The ordering of multiplications in Eq. (3.20) depends on the kinematic structure defined by $\hat{\mathbb{M}}$. This ordering can efficiently be computed for example using Dijkstra's algorithm to find the (shortest) path between two nodes in the spanning tree. Note that the marginalization of Eq. (3.20) is only valid for open kinematic trees. We generalize this procedure to arbitrary kinematic systems in Chapter 4, including kinematic systems containing kinematic loops.

In principle, the *inverse kinematic model* can be derived by applying Bayes' rule,

$$p(q_1, \ldots, q_m \mid \mathbf{x}_1, \ldots, \mathbf{x}_n) = \frac{p(q_1, \ldots, q_m)}{p(\mathbf{x}_1, \ldots, \mathbf{x}_n)} \, p(\mathbf{x}_1, \ldots, \mathbf{x}_n \mid q_1, \ldots, q_m), \tag{3.21}$$

it is in practice difficult to determine the maximum likelihood (ML) solution for the action $q_1, \ldots, q_m$. This is due to the fact that the target posture is typically not fully specified for all body parts but rather for the root part and the end effector. Thus, the Bayesian network is only constrained at both "ends", which results in a high-dimensional optimization problem.

For this reason, we resort to *differential kinematics* which uses the Jacobian to compute a configuration that moves the end effector iteratively towards the desired target pose Sciavicco and Siciliano, 2000. Since all individual functions $f_{\mathcal{M}_i}$ are continuous, the maximum likelihood estimate f from Eq. (3.20) of the forward kinematic model is continuous, too, and so the Jacobian of the forward model can be computed as

$$J_f(\mathbf{q}) = \left[\frac{\partial f(\mathbf{q})}{\partial q_1}, \ldots, \frac{\partial f(\mathbf{q})}{\partial q_m} \right]^T. \tag{3.22}$$

Given the Jacobian $J_f(\mathbf{q})$, it is straight-forward to implement a gradient descent-based algorithm that continuously minimizes the distance function and, thus, controls the manipulator towards the target pose. While such a "greedy" controller may get trapped in local minima of the distance function and might fail to plan around obstacles, it is often used in practice for manipulator control and forms the basis of many higher-level path-planning algorithms such as probabilistic road-maps or rapidly-exploring random trees (LaValle, 2006).

3.3 Failure Awareness and Life-Long Adaptation

So far, we have assumed that the kinematics of the robot remain unchanged during its life-time. It is clear, however, that in many real-world applications, the kinematics of a robot will change over the course of its life-time. This can, for example, be caused by material fatigue, wear and tear, or inaccurate repairs. This requires that the robot revises parts of its internal model over time and can discriminate between earlier and more recent observations to reason about such changes. We would like the robot to detect changes of its kinematics by testing the validity of its local models continuously. It might even be useful for the robot to maintain multiple body schema at different time scales. Consider, for example, a robot that uses an accurate pre-programmed model over a long period of time and that has the ability to learn additional models in response to kinematic changes. Such a situation is depicted in Figure 3.9. In this experiment, we changed the tool in the end effector without notifying the system. The task of the robot is to detect this change and to learn a replacement for the mismatching local model.

To deal with model changes over time, we add a time index T to the local models $\mathcal{M}^T$ to indicate this dependency. Consequently, the size of the learning problem grows exponentially in time yielding the immense upper bound of $\sum_{k=0}^{m} \binom{n}{2} \binom{m}{k} 2^{|T|}$ local models to be considered. As it is intractable to evaluate all of these local models even for small periods of time, we make three additional assumptions such that an efficient algorithm for online applications can be implemented:

1. Changes to the kinematic structure and/or kinematic properties are relatively rare events.
2. Changes happen incrementally.

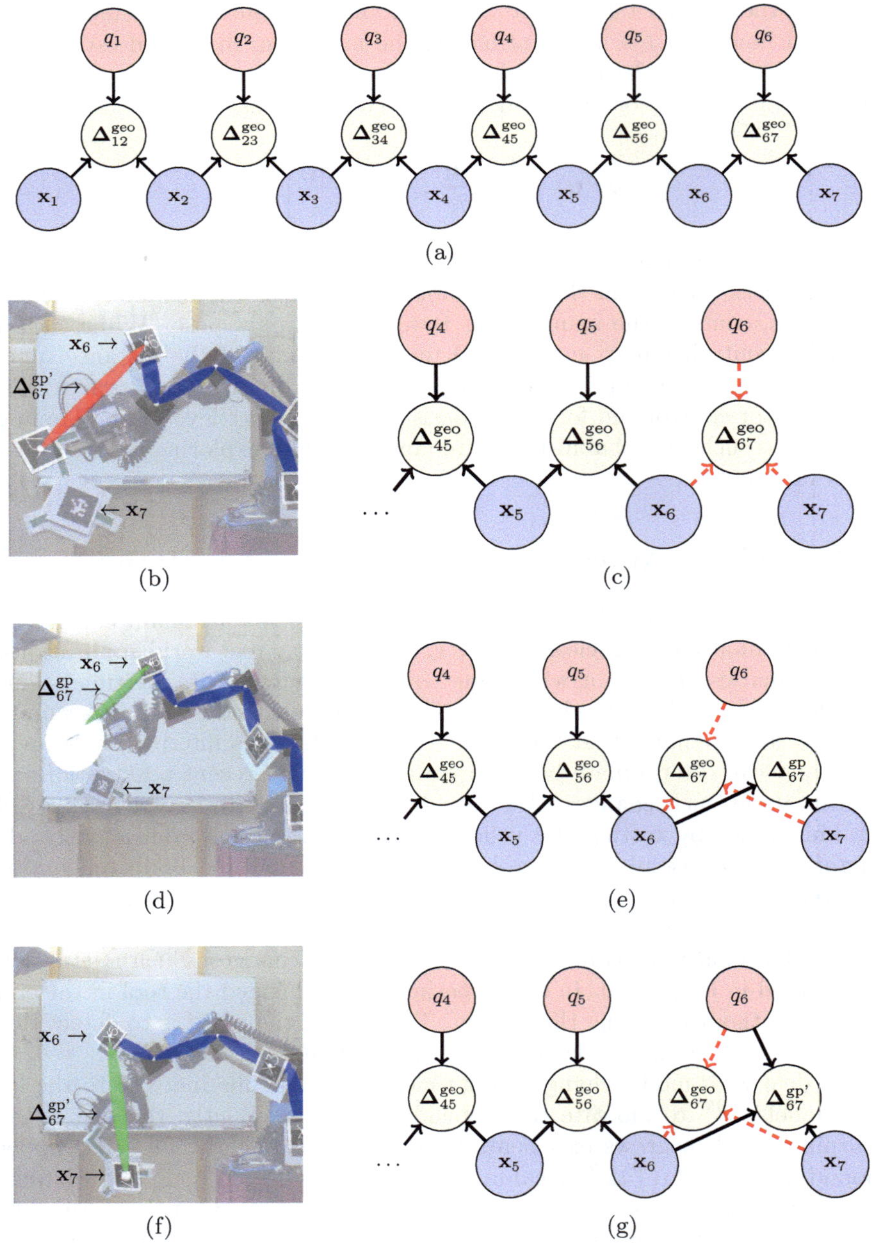

Fig. 3.9 Adaptation of the body schema during tool-use. (a) Initial body schema. (b) After a different tool is placed in the gripper, the model does not fit the observations anymore. (c) The mismatching model $\mathbf{\Delta}_{67}$ is revoked. (d)+(e) The first newly sampled model ($\mathbf{\Delta}_{67}^{gp}$) has a high uncertainty because of the missing dependency on the action signal q_6. (f)+(g) The second sampled model ($\mathbf{\Delta}_{67}^{gp'}$) is a suitable replacement.

3. Whatever local models were useful in the past, it is likely that similar – or even the same – local models will be useful in the future.

Due to the first assumption, we do not have to re-learn the local models continuously and re-optimize the network, but rather it is sufficient to monitor the data likelihood of the models until one of them is not evaluated as being *valid* any more. In this case, the second assumption states that the network cannot change completely at a given time step, but that we can recover the new structure by exchanging non-valid local models by re-learned ones individually. Furthermore, according to our third assumption, it is reasonable to begin the search for new models with those that are similar to previously useful models, i.e., to keep a history of successful local models and to start searching within this history before learning new models from scratch.

We incorporate these assumptions into an integrated system that is able to learn a body schema from scratch and to exchange local models at a later stage whenever a misfit is detected. For rating and ordering alternative local models, we consider the *structural proximity* $d_{\text{DBN}}(\mathcal{M}_1, \mathcal{M}_2)$ of two local models which we define as the ratio of shared nodes in the Bayesian network. This way, models that depend on a similar set of variables are given preference in the search. We now present an experimental evaluation of the integrated system in simulation and on two real robotic manipulators.

3.4 Experiments

We tested our approach in a series of experiments on a real robot as well as in simulation. The goal of our experiments was to verify that

1. the robot is able to learn its kinematic structure and individual transformation functions,
2. subsequent changes to the robot's body are detected reliably (blocked joints/deformations),
3. the body schema is updated automatically without human intervention, and
4. the resulting model allows for accurate prediction and control.

The two real robots used to carry out the experiments were equipped with a 2-DOF and with a 7-DOF manipulator, respectively, composed of Schunk PowerCube modules (see Figure 3.2). We compare the learned kinematic model with a carefully hand-tuned model that uses the joint encoder measurements for predicting the current pose. Note that our approach uses in contrast only the actions and not proprioception for learning the model and predicting the pose. Visual perception was implemented using a Sony DFW-SX900 FireWire camera at a resolution of 1280x960 pixels. Seven black-and-white markers were attached to the joints of the robot and the ARToolkit vision module (Fiala, 2005) was used to continuously estimate their 3D poses. The standard deviation of the camera noise was measured to $\sigma_{\text{markers}} = 0.044\,\text{m}$

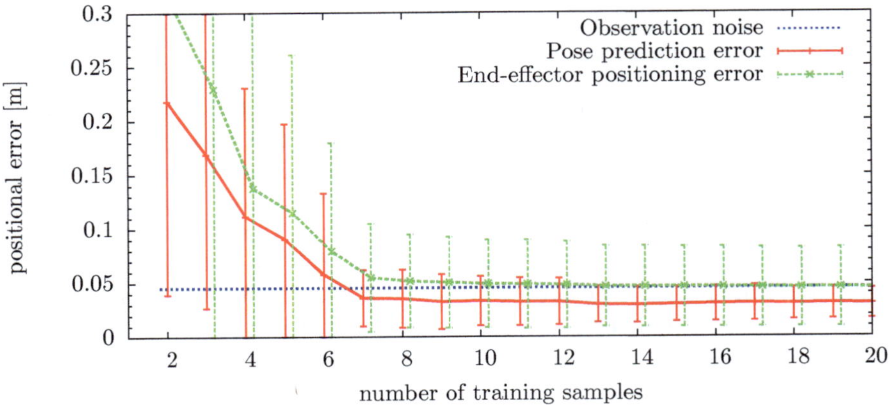

Fig. 3.10 Pose prediction and end-effector positioning errors of our model learning approach evaluated on a real 2-DOF manipulation robot.

in 3D space, which is acceptable considering that the camera was approximately located two meters away from the robot. The prediction errors and error bars reported in the following were evaluated using independent test sets $\mathcal{D}_{\text{testing}}$ with 15 data samples.

3.4.1 Evaluation of Model Accuracy

To quantitatively evaluate the accuracy of the kinematic models learned from scratch as well as the convergence behavior of our learning approach, we generated random action sequences and analyzed the intermediate models using the 2-DOF robot of which the kinematic model is perfectly known. Figure 3.10 gives the absolute errors of prediction and control after certain numbers of observations have been processed. For a reference, we also give the average observation noise, i.e., the absolute localization errors of the visual markers. As can be seen from the diagram, the body schema converges robustly within the first 10 observations. After about 15 training samples, the accuracy of the predicted body part positions becomes even higher than the accuracy of the direct observations. The latter is a remarkable result as it means that, although all local models are learned from noisy observations, the system is able to "blindly" estimate its pose more accurately than immediate perception. The figure also gives the accuracy when the robot is using the learned model to control its position. Here, we used an additional marker to define the target location of the end effector. We learned the full body schema from scratch as in the previous experiment and used the gradient-based control algorithm to bring the end effector to the desired target location. The average positioning error is in the order of the perception noise (approximately 0.050 m, see Figure 3.10), which is slightly higher than the prediction error alone.

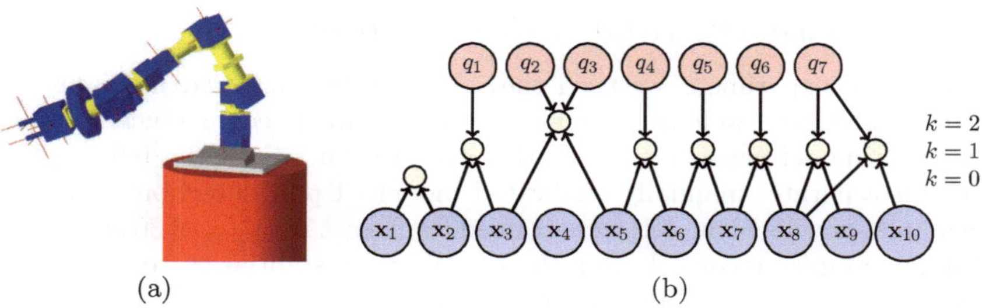

Fig. 3.11 Experiment with a simulated 7-DOF-manipulator consisting of 10 body parts. Body part $\mathbf{x}_4$ was occluded and, thus, never observed. (a) Picture of the simulated robot. (b) After 10 training samples, the Bayesian network has converged to the correct kinematic structure.

The second experiment was carried out on a 2-DOF robot of similar size in simulation. Therein, we analyzed the convergence behavior of the local models with respect to the training size in the absence of observation noise. We evaluated the accuracy of the learned models on independently drawn test sets. Here, we found that the accuracy was on average below 0.002 m and 1° after 20 training samples, and below 0.001 m and 0.2° after 100 training samples. This shows that the underlying Gaussian process regression models can approximate the kinematic function arbitrarily well, given that enough training data is available.

Further, we evaluated our algorithm on a simulated 7-DOF manipulator consisting of 10 body parts, to verify that our approach also scales to larger manipulators. The total length of the simulated manipulator was 1.300 m. The manipulator has been assembled as follows (see Figure 3.11):

- Body parts $\mathbf{x}_1$ and $\mathbf{x}_2$ were firmly connected to each other.
- Two fingers $\mathbf{x}_9$ and $\mathbf{x}_{10}$ were mounted on the 1-DOF gripper whose configuration is given by q_7.
- The remaining body constituted a chain of visible body parts $\mathbf{x}_2, \ldots, \mathbf{x}_8$ and revolute joints $q_1, \ldots, q_6$.

The structure of the learned forward model converges after around 10 samples, similar to previous experiments. The average prediction error after around 100 samples was below 0.001 m.

With these experiments, we demonstrated that our approach is able to recover the kinematic model of several real and simulated manipulators. Furthermore, we showed that the learned models are more accurate than the observation noise in the real robot experiment and asymptotically converge towards zero error in the noise-free case. Finally, with the experiment on the simulated 10 part manipulator and 7 DOFs, we demonstrated that our approach applies also to more complex structures.

3.4.2 Recovery from a Blocked Joint

In a second experiment we used the 7-DOF robot depicted in Figure 3.2b to evaluated how well the proposed system can detect a stuck joint and repair its model accordingly. To this end, we initialized the body schema with an accurate, manually calibrated model. Upon detection of a model mismatch, new local models were trained from a set $\mathcal{D}_{\text{training}}$ of 30 consecutive training samples recorded after the model was instantiated. In order for a local model to be valid, its translational and orientational error on the test set was required to be within $3\sigma_{\mathbf{z},\text{pos}} = 0.150\,\text{m}$ and $3\sigma_{\mathbf{z},\text{orient}} = 45°$, with $\sigma_{\mathbf{z},\text{pos}}$ and $\sigma_{\mathbf{z},\text{orient}}$ the standard deviations of the positional and orientational observation noise, respectively. New local models were only sampled when no valid spanning tree could be constructed for 15 consecutive time steps. This corresponds to the time it takes to replace the data samples in the test set – depending on the visibility of the individual markers.

We generated a large sequence of random actions $\langle \mathbf{q}_1, \ldots, \mathbf{q}_t \rangle$. Before accepting a pose, we checked that these actions would not cause any (self-)collisions and that the visual markers of interest would potentially be visible on the monocular camera image. This sequence was sent to the robot and after the motion of the manipulator stopped, the observed marker poses $(\mathbf{y}_1, \ldots, \mathbf{y}_n)$ were recorded. We allowed for arbitrary motion patterns (only constrained by the geometry of the manipulator) and thus do not require full visibility of the markers. In the rare case of an anticipated or actual (self-)collision during execution, the robot stopped and the sample was rejected. Analysis of the recorded data revealed that, on average, the individual markers were visible only in 86.8 % of the images. In a second run, we blocked the end effector joint q_4 so that it could not move and again recorded a log-file. An automated test procedure was then used to evaluate the performance and robustness of our approach. For each of the 20 recorded runs, a data sequence was sampled from the log-files, consisting of 4 blocks with $N = 100$ data samples each. The first and the third block were sampled from the initial body shape, while the second and the fourth block were sampled from the log-file where the joint got blocked.

Figure 3.12a shows the absolute errors of the local models predicting the end effector pose. As expected, the prediction error of the engineered local model increases significantly after the end effector joint gets blocked at $t = 100$. After a few samples, the robot detects a mismatch in its internal model and starts to learn a new dynamic model (around $t = 130$), which quickly reaches the same accuracy as the original, engineered local model. At $t = 200$, the joint gets repaired (unblocked). Now the estimated error of the replacement model quickly increases while the estimated error of the engineered local model decreases rapidly towards its initial accuracy. Later, at $t = 300$, the joint gets blocked again in the same position, the accuracy of the previously learned replacement model increases significantly, and thus the robot can re-use this local model instead of having to learn a new one.

Table 3.1 Evaluation of the number of pose observations required until the robot can re-establish a valid kinematic model after being exposed to different types of failures. The numbers give the mean and standard deviations in 20 independent runs.

Visibility rate	Failure type	Time steps until recovery		
		first occurrence	restore/repair	second occurrence
91.9 %	Joint stuck	16.50 ± 1.20	0.45 ± 0.86	0.65 ± 1.15
79.0 %	Tool exchange	20.20 ± 1.96	11.10 ± 0.83	12.10 ± 1.64

We averaged the precision of the combined model – i.e., the engineered one fused with the one learned after having detected the failure – over 20 runs of the experiment. The results are given in Figure 3.12b. The hand-tuned initial geometrical model evaluates to an averaged error at the end effector of approximately 0.037 m. After the joint gets blocked at $t = 100$, the error in prediction increases rapidly. After $t = 115$, a single new local model gets sampled, which already is enough to bring down the overall error of the combined kinematic model to approximately 0.051 m. Training of the new local model is completed at around $t = 135$.

Later, at $t = 200$, when the joint gets un-blocked, the error estimate of the combined kinematic model increases slightly, but returns much faster to its typical accuracy: switching back to an already known local model requires less data samples than learning a new model (see Table 3.1). At $t = 300$, the same quick adaption can be observed when the joint gets blocked again.

3.4.3 Tool Use

In a third experiment, we changed the end effector link length and orientation and applied the same evaluation procedure as in the previous subsection. This was accomplished by placing a tool with an attached marker in the gripper and changing its configuration during the experiment (see Figure 3.9). After a different tool is placed in the gripper, the body schema does not fit the observations anymore. In particular, the robot identifies $\mathbf{\Delta}_{67}$ as the mismatching component and seeks for a replacement. The first newly sampled model ($\mathbf{\Delta}_{67}^{gp}$) has a high uncertainty because of the missing dependency on the action signal q_6. Accordingly, the robot samples a second model $\mathbf{\Delta}_{67}^{gp'}$ which it evaluates as a suitable replacement. As a result, the adapted body schema is again valid and the robot can position its tool accurately.

The quantitative results for a single run and the average over 20 runs of this experiment are given in Figure 3.13. After the tool gets displaced at $t = 100$, two local models have to be sampled on average to repair the kinematic model. The prediction accuracy of the whole system closely resembles the levels that were obtained in the case of the blocked joint. On average, we measured an accuracy of 0.047 m after recovery. In Table 3.1, we summarize

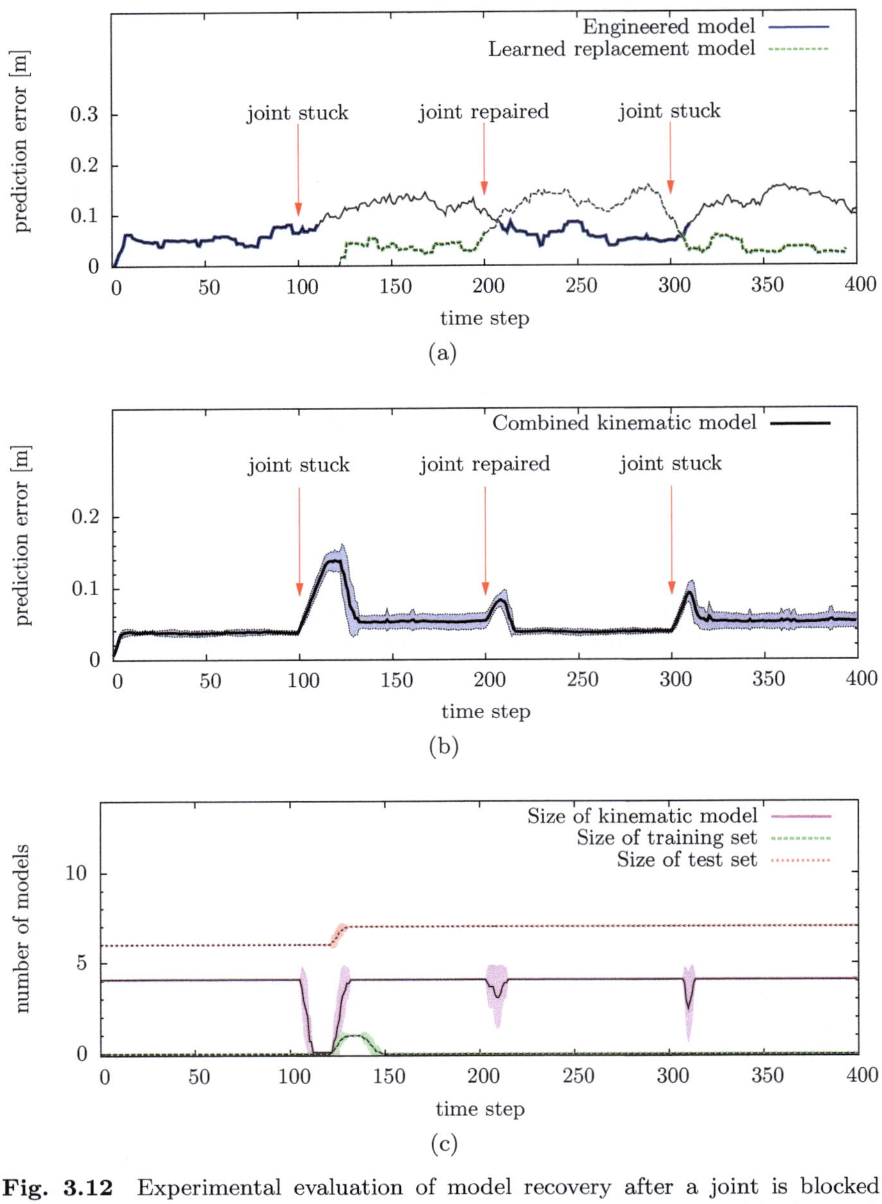

Fig. 3.12 Experimental evaluation of model recovery after a joint is blocked. (a) Prediction errors of the engineered and learned replacement model of a single run. (b) Prediction error of the combined model averaged over 20 runs. (c) Number of models in the current Bayesian network, the current training set, and the current test set. On average, our approach only needs to sample a single model before the kinematic model is restored.

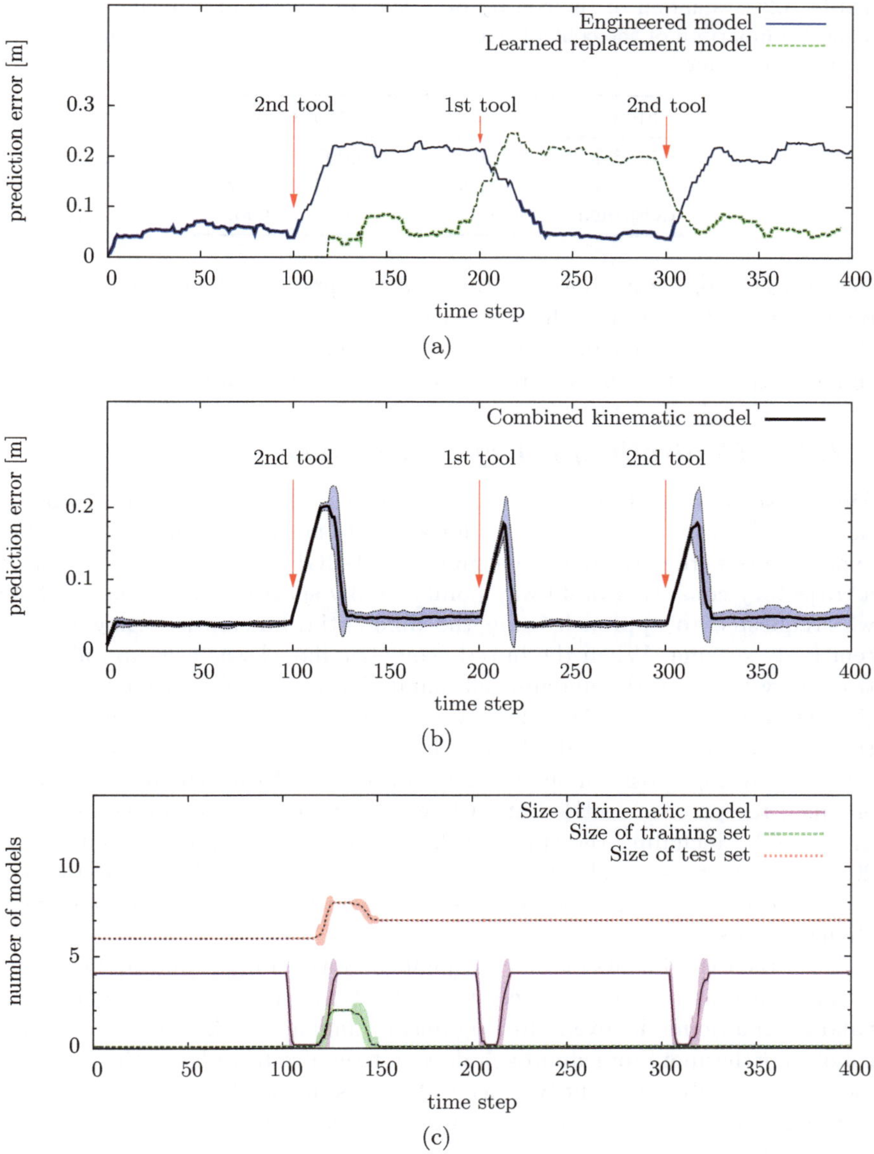

Fig. 3.13 In this experiment, the tool in the end effector of the robot was repeatedly exchanged. (a) Prediction error of the engineered and learned replacement model of a single run. (b) Prediction error of the combined model averaged over 20 runs. (c) Evolution of models being trained and tested while the kinematic model gets updated. In this case, the robot samples on average two local models before the kinematic model is restored.

Table 3.2 Evaluation of the control of a deformed robot in simulation. Experimental comparison of the control error while following a trajectory in the presence of hardware failures.

Shape	Strategy	Control error [m]
initial	static	0.007 ± 0.011
deformed	static	0.189 ± 0.028
deformed	adaptive	0.015 ± 0.002

the averaged recovery times for this and the previous experiment. As can be seen from the results, the system recovers from a blocked joint quicker than from a deformed link, and recalling a previously successful model is significantly faster than learning a new model from scratch.

3.4.4 Controlling a Deformed Robot

Finally, we performed a series of experiments to verify that dynamically maintained body schemata can be used for accurate positioning and control. The experiments were performed on a simulated 4-DOF manipulator. We defined a trajectory consisting of 30 way points (in 3D space) that the manipulator was requested to approach using the differential kinematics using its current body schema. When the initial geometric model was used to follow the trajectory by using the undamaged manipulator, the robot achieved a positioning accuracy of 0.007 m. After we had deformed the middle link by 45°, the manipulator with a static body schema was significantly off course, leading to an average positioning accuracy of 0.189 m. With dynamic adaptation enabled, the precision settled at 0.015 m. These results are also summarized in Table 3.2 including the standard deviations of the errors computed over 20 independent runs. The results show that dynamic model adaptation enables a robot to maintain a high positioning accuracy even after substantial changes to its kinematics.

With the experiments on blocked joints, deformed links, and tool changes, we showed that robots equipped with our approach are able to maintain a valid kinematic model even after significant damage or changes occur to the robot. Furthermore, our approach does not require to re-learn the complete model, but is able to identify inaccurate parts of the Bayesian network and to replace these efficiently using a suitable search heuristic. With our experiments on controlling a deformed robot, we demonstrated that a robot using our approach stays operational after link deformations and hardware failures and thus requires less human supervision.

3.5 Related Work and Discussion

The central idea of our approach is to represent the kinematic model as a probabilistic Bayesian network whose vertices correspond to body parts and action signals and whose edges encode the local kinematic models. Dearden and Demiris (2005) enabled a robot to learn a Bayesian network that relates action commands to the visual motion of its gripper. In comparison to our work, the problem considered by Dearden and Demiris is much simpler as they deal only with two body parts observed in two-dimensional camera images. As a result, their model does not provide a three-dimensional kinematic model of the manipulator.

Kuipers et al. provided with the "spatial semantic hierarchy" (SSH) a set of concepts on representing and learning sensor-motor maps for robots at different abstraction levels. Their work is inspired by the concept of human cognitive maps (Kuipers and Byun(1988); Kuipers et al.(2000); Remolina and Kuipers(2004)). The general idea is to learn mappings that relate sensor input to motor commands and that enable a robot, for example, to follow trajectories without any prior knowledge. A different approach has been presented by Kolter and Ng (2007) who applied dimensionality reduction to find a suitable subspace in order to learn a walking gait for a four-legged robot. Another instance of approaches based on dimensionality reduction is given by the work of Grimes et al. (2006) who employed principal component analysis in conjunction with Gaussian process regression to learn walking gaits for a humanoid robot. Yoshikawa et al. (2004a) used Hebbian networks to discover the body schema from self-occlusion and self-touching sensations, and learned classifiers for body/non-body discrimination from visual data (Yoshikawa et al., 2004b). By combining the sensor data across multiple modalities such as visual, proprioceptive and tactile sensor data, Sawa et al. (2007) enabled a robot to infer the Jacobian even for invisible hand positions.

Other approaches model the (inverse) kinematic function directly as a high-dimensional regression problem. For example, Natale (2004) used neural feed-forward networks to learn reaching movements, Gaskett and Cheng (2003) proposed self-organizing maps to coordinate hand-eye movements, and Kumar et al. (2010) employed radial basis function networks (RBFs) to learn the local mappings between configurations and end effector poses. Recurrent neural networks have also been used to learn the kinematics and dynamics of manipulation robots (Reinhart and Steil, 2008; Rolf et al., 2009). As no global inverse kinematic function exists for redundant kinematic chains, D'Souza et al. (2001) estimated the inverse kinematic function locally from observed data. As the required number of training samples increases exponentially with the degrees of freedom of the robot, Lopes and Santos-Victor (2005) proposed to learn the kinematic function incrementally, first by moving only the shoulder and elbow joints and, subsequently, the hand. Other approaches aim to reduce the number of training samples required to learn an accurate kinematic model. Martinez-Cantin et al. (2010) showed how active learning can be used to reduce the number of required training samples, by actively choosing joint configurations that maximize the expected information gain. Angulo and Torras (2005) approached this problem by splitting the manipulator into two or more virtual robots. However, Angulo and Torras assumed that a suitable decomposition of the manipulator is known beforehand, and thus, did not tackle the problem of learning the kinematic structure.

The approach presented in this article is also related to the problem of parameter optimization, which can be understood as a sub-problem of body schema learning. When the kinematic model is given in a parametric form, the parameters can be optimized efficiently with respect to an error measure (Gatla et al., 2007; Pradeep et al., 2010; He et al., 2010) or the data likelihood (Roy and Thrun, 1999). Hersch et al. (2008) showed that parameter optimization can also be used to adapt the body schema during tool-use, for example, to estimate the tool position and orientation. Martinez-Cantin et al. (2010) extended this approach to active learning, i.e., they generated observation actions that maximize the expected information gain. Such methods can also be used to identify the dynamic parameters such as the center of mass, the moments of inertia, etc. Ting et al. (2006), for example, presented a Bayesian approach for estimating these parameters on two different manipulation robots. In principle, these methods could be applied after our approach has bootstrapped the kinematic model, in order to refine or augment the model and achieve a faster convergence. Genetic algorithms can also be used for parameter optimization given a suitable parametrization of the kinematic model space. Bongard et al. (2006a,b) described a robotic system that continuously learns its own structure from actuation-sensation relationships. Their system generates new structure hypotheses using stochastic optimization, which are validated by generating actions and by analyzing the following sensory input. In a more general context, Bongard and Lipson (2007) studied structure learning in arbitrary nonlinear systems using similar mechanisms.

In contrast to all of the approaches described above, our approach learns both the structure as well as the functional mappings for the individual building blocks of the body schema. Furthermore, it does not require an explicit parametrization of the body schema, and the representation in form of a Bayesian network allows a robot to quickly revise its structure and to replace invalidated local models on-the-fly. Recently, Hoffmann et al. (2010) published a comprehensive review on body schema learning in robotics which includes a detailed discussion of our work.

3.6 Summary

In this chapter, we presented a novel approach to body schema learning for manipulation robots. Our first contribution is to represent the kinematics of a manipulation robot as a Bayesian network. Marginalization in the Bayes net corresponds to forward and inverse kinematics, depending on which variables are marginalized. The second contribution is an efficient algorithm for finding the kinematic model from observations. We continuously learn a large set of conditional density functions (or local models) using nonparametric regression for different hypotheses of the network structure. Given this set of models, we search for the arrangement that best explains the full system. Our approach recovers the kinematic structure by finding the minimum spanning tree in the set of possible models. To the best of our knowledge, this is the first time that Bayesian models of such complex kinematic systems have been learned from scratch using visual self-observation. In experiments carried out with real manipulation robots and in simulation, we demonstrated that our system is able to deal with missing and noisy observations, operates in full 3D space, and allows a robot to robustly control its end effector even in the presence of hardware failures. With our approach, we contribute an innovative solution that increases the dependability and accuracy of manipulation robots that operate over extended periods of time without the supervision of an expert.

Chapter 4
Learning Kinematic Models of Articulated Objects

Service robots operating in domestic environments are typically faced with a variety of objects they have to deal with to fulfill their tasks. Some of these objects are articulated such as cabinet doors and drawers, or room and garage doors. The ability to deal with such *articulated objects* is relevant for service robots, as, for example, they need to open doors when navigating between rooms and to open cabinets to pick up objects in fetch-and-carry applications.

Although the problem of operating articulated objects has been investigated by many researchers (Jain and Kemp, 2009a; Klingbeil et al., 2009; Meeussen et al., 2010; Wieland et al., 2009), most of these approaches are either entirely model-free or assume substantial prior knowledge about the model and its parameters. Whereas model-free approaches aim at releasing designers from providing any a-priori model information, knowledge about objects and their kinematic properties supports the robot in state estimation, motion prediction, and planning. As large variations exist in the properties of articulated objects in domestic environments, it is difficult to equip a robot with appropriate models for all these objects.

This chapter presents a complete probabilistic framework that enables robots to learn kinematic models of articulated objects from observations of their motion. We combine parametric and nonparametric models consistently and utilize the advantages of both methods. As a result of our approach, a robot can robustly operate articulated objects in unstructured environments. As an illustrating example, consider Figure 4.1 where a mobile manipulation robot interacts with various articulated objects in a kitchen environment, learns their kinematic properties, and infers their kinematic structure.

In this chapter, we generalize our approach on kinematic model learning from the previous chapter to articulated objects. As the configuration of articulated objects is in general not directly observable, we treat the configuration as a *latent variable* in our model. Furthermore, we add parametric models to increase the robustness of model estimation for prismatic and revolute links while we keep Gaussian process models as a flexible solution for all other links. We apply Bayesian model selection as a consistent method to compare and rank alternative models. With our approach, a robot can estimate both the kinematic structure and the degrees of freedom of the

J. Sturm: *Approaches to Probabilistic Model Learning*, STAR 89, pp. 65–111.
DOI: 10.1007/978-3-642-37160-8_4

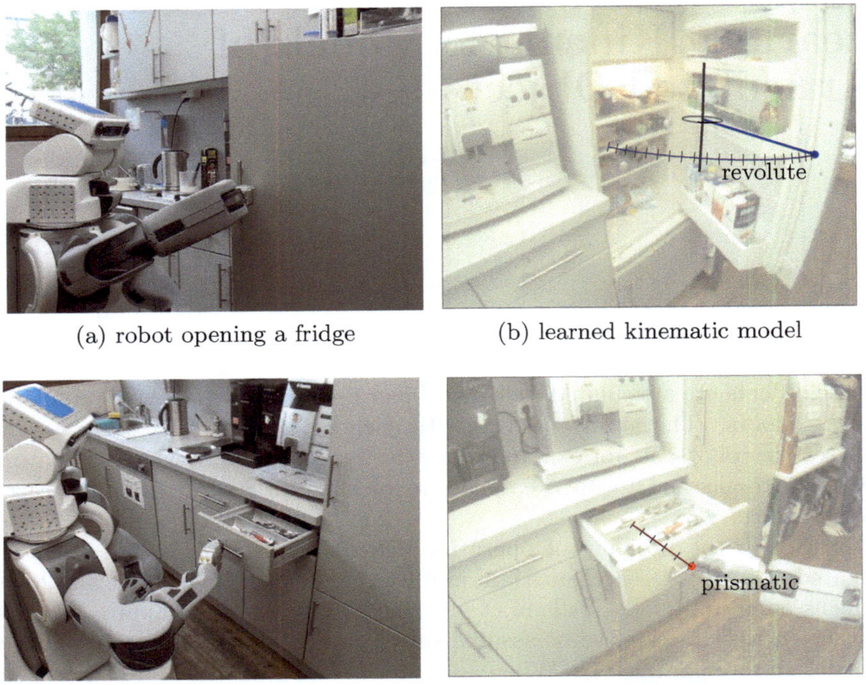

(a) robot opening a fridge (b) learned kinematic model

(c) robot opening a drawer (d) learned kinematic model

Fig. 4.1 A service robot learns kinematic models of articulated objects in a kitchen environment by interacting with them.

articulated object. Furthermore, our framework can deal with closed kinematic chains and allows the incorporation of prior knowledge during model learning. In our experiments on real robots and in simulation, we demonstrate that robots using our approach can learn accurate kinematic models of various articulated objects, operate them reliably, and improve over time by exploiting previous experiences.

This chapter is organized as follows. In Section 4.1, we introduce our unified framework for modeling the kinematics of articulated objects. In Section 4.2, we present several extensions including the exploitation of prior information, kinematic loops, and the estimation of degrees of freedom. In Section 4.3, we describe different options to perceive and control the motion of articulated objects. We analyze our approach in an extensive set of experiments both in simulation and on real robots and report our results in Section 4.4. Finally, we conclude the chapter with a discussion of related work in Section 4.5.

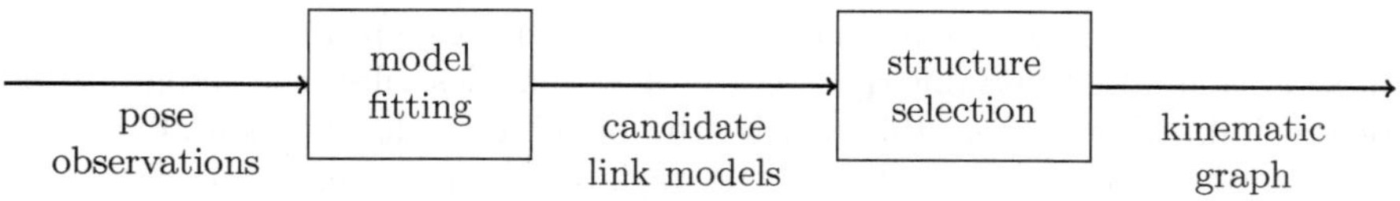

Fig. 4.2 Schematic overview of our approach.

4.1 Unified Framework for Learning Kinematic Models

We define an articulated object to consist of two or more object parts with one or more passively actuated mechanical links between them. These links constrain the motion between the parts, for example, the hinge of a door constrains the door to move on an arc, and the shaft of a drawer constrains the drawer to move on a line segment. The simplest articulated object consists of two rigid parts with one mechanical link. More complex objects may consist of several articulated parts, like a door with a door handle, or a car with several doors, windows, and wheels.

Figure 4.2 gives a high-level overview of the proposed system. A robot observes the pose of an articulated object being manipulated. For the relative motion of any two parts, it fits different candidate models that describe different mechanical links. From this set of candidate link models, it selects the kinematic structure that best explains the observed motion, i.e., the kinematic structure that maximizes the posterior probability.

We assume that a robot, external to the object, observes the pose of an articulated object consisting of p object parts. We denote the true pose of object part $i \in \{1, \ldots, p\}$ by a vector $\mathbf{x}_i \in SE(3)$ representing the 3D pose of that part (including position and orientation), where $SE(3) = \mathbb{R}^3 \times SO(3)$ stands for the special Euclidean group. Further, we refer to the full object pose (containing the poses of all parts) with the vector $\mathbf{x}_{1:p} = (\mathbf{x}_1, \ldots, \mathbf{x}_p)^T$. Two object parts i and j are related by their relative transformation $\boldsymbol{\Delta}_{ij} = \mathbf{x}_i \ominus \mathbf{x}_j$. We use $\oplus$ and $\ominus$ for referring to the motion composition operator and its inverse[1].

We denote a kinematic link model between two object parts i and j as $\mathcal{M}_{ij}$, and its associated parameter vector as $\boldsymbol{\theta}_{ij} \in \mathbb{R}^{k_{ij}}$, where $k_{ij} \in \mathbb{N}_0$ denotes the number of parameters of the model describing the link. A kinematic graph $G = (V_G, E_G)$ consists of a set of vertices $V_G = \{1, \ldots, p\}$ corresponding to the parts of the articulated object and a set of undirected edges $E_G \subset V_G \times V_G$ describing the kinematic link between two object parts. Furthermore, each edge (ij) has an associated kinematic link model $\mathcal{M}_{ij}$ with parameter vector $\boldsymbol{\theta}_{ij}$.

[1] For example, if the poses are represented as homogeneous matrices, i.e., $\mathbf{x}_1, \mathbf{x}_2 \in \mathbb{R}^{4\times 4}$, then these operators correspond to matrix multiplication $\mathbf{x}_1 \oplus \mathbf{x}_2 = \mathbf{x}_1\mathbf{x}_2$ and inverse multiplication $\mathbf{x}_1 \ominus \mathbf{x}_2 = (\mathbf{x}_1)^{-1}\mathbf{x}_2$, respectively.

All kinematic link models that we consider here (except for the trivial rigid link) have a latent variable $\mathbf{q}_{ij} \in C_{ij} \subset \mathbb{R}^{d_{ij}}$ that describes the configuration of the link. For a door, this can be the opening angle. C_{ij} stands for the configuration space of the link. The variable d_{ij} represents the *degrees of freedom* (DOFs) of the mechanical link between the two parts.

While the object is being articulated, the robot observes the object pose; we denote the n-th pose observation of object part i as $\mathbf{y}_i^n$. Correspondingly, we denote the n-th pose observation of all parts as $\mathbf{y}_{1:p}^n$ and a sequence of n pose observations as $\mathcal{D}_\mathbf{y} = \langle \mathbf{y}_{1:p}^1, \ldots, \mathbf{y}_{1:p}^n \rangle$. Further, we will refer to $\mathcal{D}_{\mathbf{z}_{ij}} = \langle \mathbf{z}_{ij}^1, \ldots, \mathbf{z}_{ij}^n \rangle$ as the sequence of relative transformations $\mathbf{z}_{ij} = \mathbf{y}_i \ominus \mathbf{y}_j$ that the robot has observed so far for the edge (ij).

Figure 4.3a depicts a graphical model of an articulated object consisting of two parts using the plate notation. The nodes inside the rectangle are copied for n times, i.e., for each time step t in which the object is observed. In each of these time steps, the articulated object takes a particular configuration $\mathbf{q}_{12}$ defining – together with the model and its parameters – the noise-free relative transformation $\boldsymbol{\Delta}_{12}$ between the noise-free pose of the object parts $\mathbf{x}_1$ and $\mathbf{x}_2$. From that, the robot observes the noisy poses $\mathbf{y}_1$ and $\mathbf{y}_2$ and infers from them a virtual measurement $\mathbf{z}_{12} = \mathbf{y}_1 \ominus \mathbf{y}_2$. During model learning, the robot infers from these observations the link model and link parameters $\mathcal{M}_{12}$ and $\boldsymbol{\theta}_{12}$, respectively.

A reduced version of this graphical model is depicted in Figure 4.3b. To improve readability, we leave out some nodes, i.e., the node corresponding to the relative transformation $\boldsymbol{\Delta}_{12}$ and the observation nodes $\mathbf{y}_1$, $\mathbf{y}_2$, and $\mathbf{z}_{12}$. Instead, we visualize the dependency between $\mathbf{x}_1$ and $\mathbf{x}_2$ by a direct link and label it with the corresponding model. Further, we collapse the configuration of the link into a single node corresponding to the configuration of the whole object. Finally, we refer to the *kinematic graph* as the graph that models the connectivity between object parts, as depicted in Figure 4.3c.

Problem Statement

The problem that we consider here is to find the most likely kinematic graph $\hat{G}$ given a sequence of pose observations $\mathcal{D}_\mathbf{y}$ of an articulated object. In Bayesian terms, this means that we aim at finding the kinematic graph $\hat{G}$ that maximizes the posterior probability of observing the poses $\mathcal{D}_\mathbf{y}$ of the articulated object, i.e.,

$$\hat{G} = \arg\max_G p(G \mid \mathcal{D}_\mathbf{y}). \tag{4.1}$$

However, finding the global maximum of the posterior $p(G \mid \mathcal{D}_\mathbf{y})$ is difficult, because it is a highly nonlinear function over a high-dimensional parameter space consisting of discrete as well as continuous dimensions that encode both the kinematic structure and the kinematic properties.

Therefore, in this section, we consider a simplified problem. We restrict the structure space to kinematic trees only and focus on the general problem

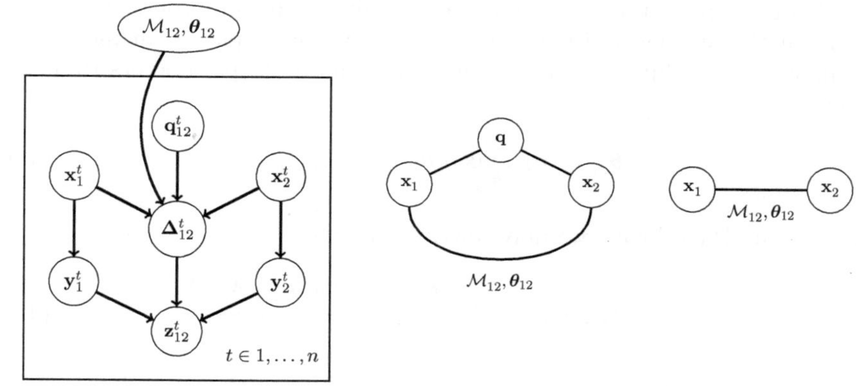

(a) full graphical model with all random variables (b) reduced graphical model (c) kinematic graph

Fig. 4.3 Three equivalent representations of kinematic models.

in Section 4.2. Kinematic trees have the property that their individual edges are independent of each other. As a result, we can estimate the link parameters independently of each other and independently of the kinematic structure. This means that for learning the local kinematic relationship between object parts i and j, only the observations of the relative transformations $\mathcal{D}_{\mathbf{z}_{ij}} = (\mathbf{z}_{ij}^1, \ldots, \mathbf{z}_{ij}^n)$ are relevant. With this insight, we can rephrase the maximization problem of Eq. (4.1) for kinematic trees now as

$$\hat{G} = \arg\max_G p(G \mid \mathcal{D}_{\mathbf{z}}) \tag{4.2}$$

$$= \arg\max_G p(\{(\mathcal{M}_{ij}, \boldsymbol{\theta}_{ij}) \mid (ij) \in E_G\} \mid \mathcal{D}_{\mathbf{z}}) \tag{4.3}$$

$$= \arg\max_G \prod_{(ij)\in E_G} p(\mathcal{M}_{ij}, \boldsymbol{\theta}_{ij} \mid \mathcal{D}_{\mathbf{z}_{ij}}). \tag{4.4}$$

The latter transformation follows from the mutual independence of the edges of kinematic trees.

The second insight in our work is that the kinematic link models representing the edges can be estimated independently from the actual structure of the kinematic tree. As a result, the problem can be solved efficiently: first, we estimate the link models of all possibles edges $(ij) \in V_G \times V_G$:

$$(\hat{\mathcal{M}}_{ij}, \hat{\boldsymbol{\theta}}_{ij}) = \arg \max_{\mathcal{M}_{ij}, \boldsymbol{\theta}_{ij}} p(\mathcal{M}_{ij}, \boldsymbol{\theta}_{ij} \mid \mathcal{D}_{\mathbf{z}_{ij}}). \tag{4.5}$$

These link models are independent of each other and independent of whether they are actually part of the kinematic structure E_G. Second, given these link models, we estimate the kinematic structure. This two-step process is also visualized in Figure 4.2.

Solving the first step, i.e., Eq. (4.5), is again a two-step process (MacKay, 2003): at the first level of inference, we assume that a particular model (for example, the revolute model) is true and estimate its parameters from the observations.

$$\hat{\boldsymbol{\theta}}_{ij} = \arg\max_{\boldsymbol{\theta}_{ij}} p(\boldsymbol{\theta}_{ij} \mid \mathcal{D}_{\mathbf{z}_{ij}}, \mathcal{M}_{ij}) \tag{4.6}$$

By applying Bayes' rule, we may rewrite this into

$$\hat{\boldsymbol{\theta}}_{ij} = \arg\max_{\boldsymbol{\theta}_{ij}} \frac{p(\mathcal{D}_{\mathbf{z}_{ij}} \mid \boldsymbol{\theta}_{ij}, \mathcal{M}_{ij})\, p(\boldsymbol{\theta}_{ij} \mid \mathcal{M}_{ij})}{p(\mathcal{D}_{\mathbf{z}_{ij}} \mid \mathcal{M}_{ij})}. \tag{4.7}$$

Here, the term $p(\boldsymbol{\theta}_{ij} \mid \mathcal{M}_{ij})$ defines the model-dependent prior over the parameter space. In our work, we assume this prior to be uniform, so that it can be dropped. Further, we can ignore the normalizing constant $p(\mathcal{D}_{\mathbf{z}_{ij}} \mid \mathcal{M}_{ij})$, as it is has no influence on the choice of the parameter vector. This results in

$$\hat{\boldsymbol{\theta}}_{ij} = \arg\max_{\boldsymbol{\theta}_{ij}} p(\mathcal{D}_{\mathbf{z}_{ij}} \mid \boldsymbol{\theta}_{ij}, \mathcal{M}_{ij}), \tag{4.8}$$

which means that fitting of a link model to the observations corresponds to the problem of maximizing the data likelihood.

At the second level of inference, we need to compare the probability of different models given the data and select the model with the highest posterior probability, i.e.,

$$\hat{\mathcal{M}}_{ij} = \arg\max_{\mathcal{M}_{ij}} p(\mathcal{M}_{ij} \mid \mathcal{D}_{\mathbf{z}_{ij}}) \tag{4.9}$$

$$= \arg\max_{\mathcal{M}_{ij}} \int p(\mathcal{M}_{ij}, \boldsymbol{\theta}_{ij} \mid \mathcal{D}_{\mathbf{z}_{ij}})\, \mathrm{d}\boldsymbol{\theta}_{ij}. \tag{4.10}$$

In general, computing the exact posterior probability of a model is difficult and, therefore, we use in our work the Bayesian information criterion (BIC) for selecting the best model according to Eq. (4.10).

As a result of this inference, we obtain for each edge $(ij) \in V_G \times V_G$ a model $\hat{\mathcal{M}}_{ij}$ with parameter vector $\hat{\boldsymbol{\theta}}_{ij}$, that best describes the motions in $\mathcal{D}_{\mathbf{z}_{ij}}$ observed between these two parts. We denote this set of all possible link models with

$$\hat{\mathbb{M}} = \{(\hat{\mathcal{M}}_{ij}, \hat{\boldsymbol{\theta}}_{ij}) \mid (ij) \in V_G \times V_G\}. \tag{4.11}$$

Given a maximum-likelihood model estimate for each link, we can now efficiently estimate the kinematic structure $E_G \subset V_G \times V_G$. Our goal is to find the subset that maximizes the posterior probability of the resulting kinematic graph, i.e.,

$$\hat{E}_G = \arg\max_{E_G} \int p(E_G, \mathbb{M} \mid \mathcal{D}_{\mathbf{z}}) \, d\mathbb{M}. \tag{4.12}$$

We solve the equation again by maximizing the BIC over all possible structures E_G using the maximum-likelihood estimates of the models $\hat{\mathbb{M}}$ to approximate the integral.

With this factorization, we provide an efficient way for finding the kinematic model of a kinematic object: (1) we fit all models to the data, (2) select the best model for each link, and (3) find the kinematic structure of the whole articulated object. In the following, we explain how to solve the model fitting problem of Eq. (4.8) and the model selection problem of Eq. (4.10) efficiently and robustly from noisy observations. In Section 4.1.3, we present a solution how to efficiently solve Eq. (4.12) given the link models. Finally in Section 4.2, we generalize our approach to general kinematic graphs including structures with kinematic loops.

Observation Model

We start by introducing our observation model. For now, we consider simple articulated objects consisting of only $p = 2$ rigid parts and drop the ij indices to increase readability. We consider the case that the robot has observed a sequence of n relative transformations $\mathcal{D}_{\mathbf{z}} = < \mathbf{z}^1, \ldots, \mathbf{z}^n >$ between two adjacent rigid parts of an articulated object. We assume the presence of Gaussian noise in each of the measurements $\mathbf{z}$ with zero mean and covariance $\Sigma_{\mathbf{z}} \in \mathbb{R}^{6\times 6}$.

Further, we assume that a small fraction of the observations are real outliers that cannot be explained by the Gaussian noise assumption alone. These outliers may be the result of poor perception, bad data association, or other sensor failures that are hard to be modeled explicitly. As these outliers are not related to the true value of $\mathbf{\Delta} = \mathbf{x}_1 \ominus \mathbf{x}_2$ at all, we assume that they come from a uniform prior distribution. One can think of this as a latent variable $v \in \{0, 1\}$ indicating whether an observation is an inlier ($v = 1$) or an outlier ($v = 0$). Further, we denote with γ the probability of drawing an outlier, i.e., $p(v = 0) = \gamma$. Our full observation model then becomes

$$\mathbf{z} \sim \begin{cases} \mathbf{\Delta} + \mathcal{N}(0, \Sigma_{\mathbf{z}}) & \text{if } v = 1 \\ \mathcal{U} & \text{if } v = 0 \end{cases}. \tag{4.13}$$

The resulting data likelihood for a single observation $\mathbf{z}$ thus is a mixture of a Gaussian and a uniform distribution with mixing constant γ:

$$p(\mathbf{z} \mid \mathbf{\Delta}, \gamma) = (1 - \gamma) p(\mathbf{z} \mid v = 1) + \gamma p(\mathbf{z} \mid v = 0). \tag{4.14}$$

Note that in general neither the true transformation $\mathbf{\Delta}$ nor the outlier ratio γ are directly observable and thus need to be estimated from the data. To compare models with different outlier ratios, we assume a prior on the outlier ratio of $p(\gamma) \propto \exp(-w\gamma)$ with w being a weighting constant and, thereby, favor models with fewer outliers over models with more outliers. The resulting data likelihood of an observation $\mathbf{z}$ given its true value $\boldsymbol{\Delta}$ thus becomes:

$$p(\mathbf{z} \mid \boldsymbol{\Delta}) = p(\mathbf{z} \mid \boldsymbol{\Delta}, \gamma)p(\gamma). \tag{4.15}$$

Candidate Models

When considering the set of objects relevant for a service robot, one quickly realizes that the joints in many articulated objects belong to a few generic classes. In particular, revolute and prismatic joints occur frequently, although a few objects are composed of other mechanical linkages, for example spherical joints, screws, or two-bar links. Examples of revolute joints include doors, door handles, and windows. This also includes the doors of dishwashers, microwave ovens or washing machines. Examples of articulated objects with prismatic joints include drawers, sliding doors, and window blinds. However, there are also objects that have different mechanical linkages such as garage doors or two-bar desk lamps. This motivates the use of a set of candidate models, that are well suited for describing the kinematic properties of a particular class of articulated links. Our candidate set consists of parametric and nonparametric models, in particular, it includes a model for revolute joints ($\mathcal{M}^{\text{revolute}}$), for prismatic joints ($\mathcal{M}^{\text{prismatic}}$), and rigid transformations ($\mathcal{M}^{\text{rigid}}$). Additionally, there may be articulations that do not correspond to these standard motions, for which we consider a parameter-free model ($\mathcal{M}^{\text{GP}}$). This model uses a combination of dimensionality reduction and Gaussian process regression to represent arbitrary joints.

In our framework, a model class defines the conditional probability distributions $p(\boldsymbol{\Delta} \mid \mathbf{q}, \mathcal{M}, \boldsymbol{\theta})$ and $p(\mathbf{q} \mid \boldsymbol{\Delta}, \mathcal{M}, \boldsymbol{\theta})$ by means of a *forward kinematic function* $f_{\mathcal{M},\boldsymbol{\theta}}(\mathbf{q}) = \boldsymbol{\Delta}$ and the *inverse kinematic function* $f^{-1}_{\mathcal{M},\boldsymbol{\theta}}(\mathbf{z}) = \mathbf{q}$. This means that we assume that our link models are deterministic, and we attribute all noise to measurement noise in the observations of the object parts, i.e., by means of the observation model $p(\boldsymbol{\Delta} \mid \mathbf{z})$ defined in Section 4.1.

Since we have no prior information about the nature of the connection between the two rigid parts, we do not aim to fit only a single model. Instead, our approach is to fit all candidate models to the observed data and subsequently, from this set, select the best model.

4.1.1 Model Fitting

For estimating the parameters of any of the above-mentioned models, we need to find a parameter vector $\boldsymbol{\theta} \in \mathbb{R}^k$ that maximizes the data likelihood given the model, i.e.,

$$\hat{\boldsymbol{\theta}} = \arg\max_{\boldsymbol{\theta}} p(\mathcal{D}_{\mathbf{z}} \mid \mathcal{M}, \boldsymbol{\theta}). \tag{4.16}$$

In the presence of noise and outliers, finding the right parameter vector $\boldsymbol{\theta}$ that minimizes Eq. (4.16) is not trivial, as least squares estimation is sensitive to

Table 4.1 Overview of the four candidate models used in our approach.

candidate model $\mathcal{M}$	DOFs d	parameters k
rigid model	0	6
prismatic model	1	9
revolute model	1	12
Gaussian process model	$1, \ldots, 5$	$1 + d + 6n$

outliers and thus not sufficient given our observation model. Therefore, we use the MLESAC (maximum likelihood consensus) algorithm as introduced by Torr and Zisserman (2000). We estimate the initial kinematic parameters from a minimal set of samples randomly drawn from the observation sequence that we subsequently refine using nonlinear optimization of the data likelihood.

The MLESAC procedure for a model $\mathcal{M}$ works as follows: First, we generate a guess for the parameter vector $\hat{\boldsymbol{\theta}}$ in Eq. (4.16) from a minimal set of samples from $\mathcal{D}_{\mathbf{z}}$. For this guess, we compute the data likelihood of the whole observation sequence $\mathcal{D}_{\mathbf{z}}$ as the product over all data

$$p(\mathcal{D}_{\mathbf{z}} \mid \mathcal{M}, \hat{\boldsymbol{\theta}}) = \prod_{t=1}^{n} p(\mathbf{z}^t \mid \mathcal{M}, \hat{\boldsymbol{\theta}}). \tag{4.17}$$

We repeat this sampling step for a fixed number of iterations and finally select the parameter vector maximizing Eq. (4.17). On this initial guess, we apply nonlinear optimization on the data likelihood to refine the parameter vector using Broyden-Fletcher-Goldfarb-Shanno (BFGS) optimization, which is a quasi-Newton method for function maximization. During the maximization of the data likelihood, MLESAC iteratively also estimates the outlier ratio γ, using the Expectation Maximization algorithm.

In the following, we show for each of our link models how to (1) estimate the parameter vector $\boldsymbol{\theta}$ from a minimal sample set of observations, (2) estimate a transformation $\mathbf{z}$ given a configuration $\mathbf{q}$, and (3) estimate the configuration $\mathbf{q}$ given a transformation $\mathbf{z}$. A brief summary of the model candidates presented in this section is given in Table 4.1.

Rigid Model

We parametrize a rigid link by a fixed relative transformation between two object parts. Thus, the parameter vector $\boldsymbol{\theta}$ has $k = 6$ dimensions. During the sampling consensus step, we draw a single observation $\mathbf{z}$ from the training data $\mathcal{D}_{\mathbf{z}}$ that gives us an initial guess for the parameter vector $\hat{\boldsymbol{\theta}}$. This parameter vector thus corresponds to the estimated fixed relative transformation between the two parts. For the rigid transformation model, the forward kinematics function for the rigid model equals the parameter vector as it corresponds to the estimated fixed relative transform between the two parts:

$$f_{\mathcal{M}^{\text{rigid}},\boldsymbol{\theta}}(\mathbf{q}) = \boldsymbol{\theta}. \tag{4.18}$$

As the rigid model has zero DOFs ($d = 0$), an inverse kinematic function is not needed.

Prismatic Model

Prismatic joints move along a single axis and thus have a one-dimensional configuration space. The prismatic model describes a translation along a vector of unit length $\mathbf{e} \in \mathbb{R}^3$ relative to some fixed origin $\mathbf{a} \in SE(3)$. This results in a parameter vector $\boldsymbol{\theta} = (\mathbf{a}; \mathbf{e})$ with $k = 9$ dimensions.

To estimate these parameters, we sample two observations from the training data. For this, we pick the transformation of the first sample as the origin $\mathbf{a}$ and the normalized vector between them as the prismatic axis $\mathbf{e}$.

A configuration $q \in \mathbb{R}$ then encodes the distance from the origin $\mathbf{a}$ along the direction of motion $\mathbf{e}$. The forward kinematics function for the prismatic model $\mathcal{M}^{\text{prismatic}}$ is

$$f_{\mathcal{M}^{\text{prismatic}},\boldsymbol{\theta}}(\mathbf{q}) = \mathbf{a} \oplus q\mathbf{e}. \tag{4.19}$$

Let $trans(\cdot)$ be the function that removes all rotational components. The inverse kinematic function then becomes

$$f^{-1}_{\mathcal{M}^{\text{prismatic}},\boldsymbol{\theta}}(\mathbf{z}) = \mathbf{e}^T trans(\mathbf{a} \ominus \mathbf{z}). \tag{4.20}$$

Revolute Model

The revolute model describes the motion of a revolute joint, i.e., a one-dimensional motion along a circular arc. We parametrize this model by the center of rotation $\mathbf{c} \in SE(3)$ and a rigid transformation $\mathbf{r} \in SE(3)$ from the center to the moving part. This yields a parameter vector $\boldsymbol{\theta} = (\mathbf{c}; \mathbf{r})$ with $k = 12$ dimensions.

For the revolute model, we sample three observations $\mathbf{z}^i$, $\mathbf{z}^j$ and $\mathbf{z}^k$ from the training data. First, we estimate the plane spanned by these three points; its plane normal then corresponds to the rotation axis. Second, we compute the circle center as the intersection of the perpendicular lines of the line segments between the three observations. Together with the rotation axis, this gives us the center of rotation $\mathbf{c}$. Finally, we estimate the rigid transformation $\mathbf{r}$ of the circle from the first sample.

For the forward kinematic function, we obtain for revolute links

$$f_{\mathcal{M}^{\text{revolute}},\boldsymbol{\theta}}(\mathbf{q}) = \mathbf{c} \oplus Rot_Z(\mathbf{q}) \oplus \mathbf{r}, \tag{4.21}$$

where $Rot_Z(\mathbf{q})$ denotes a rotation around the Z-axis by $\mathbf{q}$. Thus, $\mathbf{q} \in \mathbb{R}$ specifies the angle of rotation. For estimating the configuration of a revolute joint we use

$$f^{-1}_{\mathcal{M}^{\text{revolute}},\boldsymbol{\theta}}(\mathbf{z}) = Rot_Z^{-1}(\mathbf{c} \ominus (\mathbf{z} \ominus \mathbf{r})), \tag{4.22}$$

where $Rot_Z^{-1}(\cdot)$ gives the rotation around the Z-axis.

Gaussian Process Model

Although rigid transformations in combination with revolute and prismatic joints might seem at the first glance to be sufficient for a huge class of articulated objects, many real-world objects cannot be described by a single shifting or rotation axis. Examples for such objects include garage doors or desk lamps, as well as furniture whose joints have aged and became loose.

Therefore, we provide additionally a nonparametric model which is able to describe more general kinematic links. This model is based on dimensionality reduction to recover the latent configuration manifold and Gaussian process regression to learn a generative, predictive model. For example, consider the motion of two object parts described by a sequence of relative pose observations $\mathcal{D}_{\mathbf{z}}$. Depending on the DOFs d of this link, the data samples will lie on or close to a d-dimensional manifold with $1 \leq d \leq 6$ being nonlinearly embedded in $SE(3)$.

There are many different dimensionality reduction techniques such as principal component analysis (PCA) for linear manifolds, or Isomap and locally linear embedding (LLE) for nonlinear manifolds (Tenenbaum et al., 2000; Roweis and Saul, 2000). In our experiments, we used both PCA and LLE for dimensionality reduction. PCA has the advantage of being more robust against noise for near-linear manifolds, while LLE is more general and can also model strongly nonlinear manifolds.

The general idea is that we use the dimensionality reduction technique to obtain the inverse kinematics function $f^{-1}_{\mathcal{M}^{\text{GP}}} : SE(3) \to \mathbb{R}^d$. As a result, we can assign configurations to each of the observations, i.e.,

$$f^{-1}_{\mathcal{M}^{\text{GP}}}(\mathbf{z}) = \mathbf{q}. \tag{4.23}$$

These assignments of observations to configurations can now be used to learn the forward kinematics function $f_{\mathcal{M}^{\text{GP}},\boldsymbol{\theta}}(\cdot)$ from the observations. Except for linear actuators, we expect this function to be strongly nonlinear.

A flexible approach for solving such nonlinear regression problems given noisy observations are Gaussian process (GPs) models. The main feature of the Gaussian process framework is, that the observed data points are explicitly included in the model and, thus, no parametric form of $f_{\mathcal{M}^{\text{GP}}} : \mathbb{R}^d \to SE(3)$ needs to be specified. Data points can be added to a GP at any time, which facilitates incremental and online learning. For this model, we aim to learn a GP that fits the dependency

$$f_{\mathcal{M}^{\text{GP}}}(\mathbf{q}) + \boldsymbol{\epsilon} = \mathbf{z} \tag{4.24}$$

for the unknown forward model underlying the articulated link under consideration. We assume homoscedastic Gaussian noise, i.e., independent and identically distributed noise terms $\boldsymbol{\epsilon} \sim \mathcal{N}(0, \Sigma_{\mathbf{z}})$. For simplicity, we train 12 independent Gaussian processes for the free components of a homogeneous 4×4 transformation matrix similar to the approach in Chapter 3. As a consequence of this over-parametrization, the predicted transformation matrices are not necessarily valid. In practice, however, they are very close to valid transformation matrices, that can be found using ortho-normalization via singular value decomposition. In our approach, we use the standard choice for the covariance function, the squared exponential. It describes the relationship between two configurations $\mathbf{q}_i$ and $\mathbf{q}_j$ in configuration space by

$$k(\mathbf{q}_i, \mathbf{q}_j) = \sigma_f^2 \exp\left(-\frac{1}{2}(\mathbf{q}_i - \mathbf{q}_j)^T \Lambda^{-1} (\mathbf{q}_i - \mathbf{q}_j)\right), \tag{4.25}$$

where σ_f^2 is the signal variance and $\Lambda^{-1} = \operatorname{diag}(l_1, \ldots, l_d)$ is the diagonal matrix of the length-scale parameters. This results in a $(1+d)$-dimensional hyper-parameter vector $\boldsymbol{\theta} = (\sigma_f^2, l_1, \ldots, l_d)$. As GPs are data-driven, they require the full set of training data for making predictions. Therefore, we count all data samples as parameters of our model, so that the number of parameters becomes $k = (1+d)+6n$, where n is the number of observations. We refer the interested reader to the text book by Rasmussen and Williams (2006) for more details about GP regression.

Note that this GP link model directly generalizes to higher-dimensional configuration spaces, i.e., with $d > 1$: after the dimensionality reduction from observations in $SE(3)$ to configurations in $\mathbb{R}^d$, we can learn a Gaussian process regression for the mapping from the configuration space $\mathbb{R}^d$ back to transformations in $SE(3)$. Our model is similar to the GPLVM model introduced by Lawrence (2005). In contrast to GPLVM, we do not optimize the latent configurations for maximizing the data likelihood as this invalidates the inverse kinematics function in Eq. (4.23). Correspondingly, the GPLVM model maps only from latent space to data space. With our approach, we can map in both directions, i.e., from pose observations to configurations, and vice versa.

4.1.2 Model Evaluation

To evaluate how well a single observation $\mathbf{z}$ is explained by a model, we have to evaluate $p(\mathbf{z} \mid \mathcal{M}, \boldsymbol{\theta})$. As the configuration is latent, i.e., not observable by the robot, we have to integrate over all possible values of $\mathbf{q}$, i.e.,

$$p(\mathbf{z} \mid \mathcal{M}, \boldsymbol{\theta}) = \int p(\mathbf{z} \mid \mathbf{q}, \mathcal{M}, \boldsymbol{\theta}) p(\mathbf{q} \mid \mathcal{M}, \boldsymbol{\theta}) \, d\mathbf{q}. \tag{4.26}$$

Under the assumption that all DOFs of the link are independent of each other and that all configuration states $\mathbf{q}$ are equally likely, we may write

$$p(\mathbf{q} \mid \mathcal{M}, \boldsymbol{\theta}) \approx n^{-d}, \tag{4.27}$$

where $n = |\mathcal{D}_\mathbf{z}|$ is the number of observations so far and thus the number of estimated configurations in the d-dimensional configuration space. With this, Eq. (4.26) can be simplified to

$$p(\mathbf{z} \mid \mathcal{M}, \boldsymbol{\theta}) \approx n^{-d} \int p(\mathbf{z} \mid \mathbf{q}, \mathcal{M}, \boldsymbol{\theta}) \, \mathrm{d}\mathbf{q}. \tag{4.28}$$

If we assume that $p(\mathbf{z} \mid \mathbf{q}, \mathcal{M}, \boldsymbol{\theta})$ is a unimodal distribution, an approximation of the integral is to evaluate it only at the estimated configuration $\hat{\mathbf{q}}$ given observation $\mathbf{z}$ using the inverse kinematics function of the model under consideration, i.e.,

$$\hat{\mathbf{q}} = f^{-1}_{\mathcal{M},\boldsymbol{\theta}}(\mathbf{z}). \tag{4.29}$$

For this configuration, we compute the expected transformation $\hat{\boldsymbol{\Delta}}$ using the forward kinematics function of the model,

$$\hat{\boldsymbol{\Delta}} = f_{\mathcal{M},\boldsymbol{\theta}}(\hat{\mathbf{q}}). \tag{4.30}$$

Given the observed transformation $\mathbf{z}$ and the expected transformation $\hat{\boldsymbol{\Delta}}$, we can now efficiently compute the data likelihood of Eq. (4.28) using the observation model from Eq. (4.15) as

$$p(\mathbf{z} \mid \mathcal{M}, \boldsymbol{\theta}) \approx n^{-d} p(\mathbf{z} \mid \hat{\boldsymbol{\Delta}}). \tag{4.31}$$

Intuitively, the approximation of the integral based on the forward and inverse kinematics model corresponds to a projection of the noisy observations onto the model. Finally, the marginal data likelihood over the whole observation sequence becomes the product over all individual likelihoods, i.e.,

$$p(\mathcal{D}_{\mathbf{z}} \mid \mathcal{M}, \boldsymbol{\theta}) = \prod_{\mathbf{z} \in \mathcal{D}_{\mathbf{z}}} p(\mathbf{z} \mid \mathcal{M}, \boldsymbol{\theta}). \tag{4.32}$$

4.1.3 Model and Structure Selection

After having fitted all model candidates to an observation sequence $\mathcal{D}_{\mathbf{z}}$, we need to select the model that best explains the data. For Bayesian model selection, this means that we need to compare the posterior probability of the models given the data, i.e.,

$$p(\mathcal{M} \mid \mathcal{D}_{\mathbf{z}}) = \int \frac{p(\mathcal{D}_{\mathbf{z}} \mid \mathcal{M}, \boldsymbol{\theta}) p(\boldsymbol{\theta} \mid \mathcal{M}) p(\mathcal{M})}{p(\mathcal{D}_{\mathbf{z}})} \, \mathrm{d}\boldsymbol{\theta}. \tag{4.33}$$

While the evaluation of the model posterior is in general difficult, it can be approximated efficiently based on the Bayesian information criterion (BIC). We denote with k the number of parameters of the current model under consideration and n the number of observations in the training data. Then, the BIC is defined as

$$\mathrm{BIC}(\mathcal{M}) = -2 \log p(\mathcal{D}_{\mathbf{z}} \mid \mathcal{M}, \hat{\boldsymbol{\theta}}) + k \log n, \tag{4.34}$$

where $\hat{\boldsymbol{\theta}}$ is the maximum likelihood parameter vector. Model selection now reduces to selecting the model that has the lowest BIC, i.e.,

$$\hat{\mathcal{M}} = \arg\min_{\mathcal{M}} \mathrm{BIC}(\mathcal{M}). \tag{4.35}$$

More details on model comparison and model selection are given in Chapter 2, and a detailed derivation of the BIC is presented in Appendix B.

Finding the Connectivity

So far, we ignored the question of connectivity and described how to evaluate and select a model $\mathcal{M}$ only for a single link between two parts of an articulated object. In this section, we extend our approach to efficiently find kinematic trees for articulated objects consisting of multiple parts.

We adopt the connectivity model from Featherstone and Orin (2008) for modeling the kinematic structure as an undirected graph $G = (V_G, E_G)$. The nodes V_G in this graph correspond to the poses of the individual object parts, while the edges E_G correspond to the links between these parts. We now reintroduce the ij-indices, i.e., use $\mathcal{D}_{\mathbf{z}_{ij}}$ to refer to the observations of link (ij) and $\mathcal{D}_{\mathbf{z}}$ to refer to the observations of the whole articulated object. $\mathcal{D}_{\mathbf{z}}$ thus contains the observations of all edges in the graph G, i.e., $\mathcal{D}_{\mathbf{z}} = \{\mathcal{D}_{\mathbf{z}_{ij}} \mid (ij) \in E_G\}$. In the previous section, we established an algorithm that fits and selects for any given edge (ij) in this graph a corresponding link model $\hat{\mathcal{M}}_{ij}$ with parameter vector $\hat{\boldsymbol{\theta}}_{ij}$. Given this, we now need to select the kinematic structure E_G, that describes which of these link models are actually present in the articulated object under consideration.

For the moment, we restrict ourselves to objects that are kinematic trees, i.e., mechanisms without kinematic loops. We consider a fully connected graph with p vertices, i.e., one vertex for each object part of the articulated object. The set of possible kinematic trees for the articulated object is now given by all spanning trees of this graph. Explicitly computing, evaluating, and reasoning with all possible kinematic trees, however, is not tractable in practice. We therefore seek to find the kinematic structure E_G that maximizes the posterior as stated previously in Eq. (4.12),

$$\hat{E}_G = \arg\max_{E_G} p(E_G \mid \mathcal{D}_{\mathbf{z}}) \tag{4.36}$$

$$= \arg\max_{E_G} p(\{(\hat{\mathcal{M}}_{ij}, \hat{\boldsymbol{\theta}}_{ij}) \mid (ij) \in E_G\} \mid \mathcal{D}_{\mathbf{z}}) \tag{4.37}$$

$$= \arg\max_{E_G} \prod_{(ij)\in E_G} p(\hat{\mathcal{M}}_{ij}, \hat{\boldsymbol{\theta}}_{ij} \mid \mathcal{D}_{\mathbf{z}}) \tag{4.38}$$

$$= \arg\max_{E_G} \sum_{(ij)\in E_G} \log p(\hat{\mathcal{M}}_{ij}, \hat{\boldsymbol{\theta}}_{ij} \mid \mathcal{D}_{\mathbf{z}}). \tag{4.39}$$

The independence assumption of the individual links for kinematic trees allows us to write the posterior of the kinematic model for the whole object in Eq. (4.37) as the product over the posteriors of the individual links in Eq. (4.38). After taking the logarithm in Eq. (4.39), the structure selection problem takes a form that can be solved efficiently. The key insight here is that the kinematic tree that maximizes Eq. (4.39) corresponds to the problem of selecting the minimum spanning tree in a fully connected graph whose edge costs correspond to the negative log posterior, i.e.,

$$\mathrm{cost}_{ij} = -\log p(\mathcal{M}_{ij}, \boldsymbol{\theta}_{ij} | \mathcal{D}_{\mathbf{z}_{ij}}). \tag{4.40}$$

These edge costs can efficiently be approximated by the BIC. The sum over these edge costs then corresponds to the negative log posterior of the kinematic tree. Therefore, the minimum spanning tree of this graph maximizes the posterior of Eq. (4.39) and thus is the solution to the maximization problem of Eq. (4.39). The minimum spanning tree can be found efficiently, i.e., in $O(p^2 \log p)$ time, for example using Prim's or Kruskal's algorithm (Cormen et al., 2001).

4.2 Framework Extensions

With the approach described so far, a robot can learn accurate kinematic models of articulated objects from observation. In the following, we consider three extensions. The first extension allows the robot to exploit prior knowledge learned during previous interactions. Second, we generalize our approach to general kinematic graphs, i.e., consider additionally objects that contain closed kinematic chains. Third, we show that estimating the DOFs of articulated objects directly follows from our approach.

Learning and Exploiting Priors

Using the approach from above, a robot always starts learning a model from scratch when it observes movements of a new articulated object. From a learning perspective, this may be seen as unsatisfactory since most articulated objects encountered in man-made environments belong to few different classes with similar parameters. For example, in a specific office or kitchen, many cabinet doors will open in the same way, i.e., have the same radius and rotation axis. Thus, a robot operating in such environments over extended periods of time can significantly boost its performance by learning priors over the space of possible articulated object models.

This section describes an extension to our approach that allows a robot to learn priors for articulated objects and a means to exploit them as early as possible while manipulating previously unseen articulated objects. Our goal is to transfer model information contained in already learned models to newly seen articulated objects. Our key idea is to identify a small set of representative models for the articulated objects and to utilize this as prior information to improve model learning when handling new objects.

To keep the notation simple, consider the case that we have previously encountered two articulated objects consisting of two parts and thus a single link only. Their observed motion is given by two observation sequences $\mathcal{D}_{\mathbf{z},1}$

and $\mathcal{D}_{\mathbf{z},2}$. The question now is whether both trajectories should be described by two distinct models $\mathcal{M}_1$ and $\mathcal{M}_2$ or by a joint model $\mathcal{M}_{1+2}$. In the first case, we can split the posterior as the two models are mutually independent, i.e.,

$$p(\mathcal{M}_1, \mathcal{M}_2 \mid \mathcal{D}_{\mathbf{z},1}, \mathcal{D}_{\mathbf{z},2}) = p(\mathcal{M}_1 \mid \mathcal{D}_{\mathbf{z},1}) p(\mathcal{M}_2 \mid \mathcal{D}_{\mathbf{z},2}). \tag{4.41}$$

In the latter case, both trajectories are explained by a single, joint model $\mathcal{M}_{1+2}$ with a parameter vector $\boldsymbol{\theta}_{1+2}$, that is estimated from the joint data $\mathcal{D}_{\mathbf{z},1} \cup \mathcal{D}_{\mathbf{z},2}$. For future reference, we denote the corresponding posterior probability as

$$p(\mathcal{M}_{1+2} \mid \mathcal{D}_{\mathbf{z},1}, \mathcal{D}_{\mathbf{z},2}). \tag{4.42}$$

To determine whether a joint model is better than two separate models by comparing the posterior probabilities from Eq. (4.41) and Eq. (4.42), one can evaluate

$$p(\mathcal{M}_{1+2} \mid \mathcal{D}_{\mathbf{z},1}, \mathcal{D}_{\mathbf{z},2}) > p(\mathcal{M}_1 \mid \mathcal{D}_{\mathbf{z},1}) p(\mathcal{M}_2 \mid \mathcal{D}_{\mathbf{z},2}). \tag{4.43}$$

This expression can efficiently be approximated with the BIC as follows. The joint model is learned from $n = n_1 + n_2$ data points, using k parameters and has a data likelihood of $L = p(\mathcal{M}_{1+2} \mid \mathcal{D}_{\mathbf{z},1}, \mathcal{D}_{\mathbf{z},2})$. In contrast, the two separate models are learned from n_1 and n_2 samples using k_1 and k_2 parameters and have data likelihoods of $L_1 = p(\mathcal{D}_{\mathbf{z},1} \mid \mathcal{M}_1)$ and $L_2 = p(\mathcal{D}_{\mathbf{z},2} \mid \mathcal{M}_2)$, respectively. Accordingly, we need to check whether

$$\mathrm{BIC}(\mathcal{M}_{1+2} \mid \mathcal{D}_{\mathbf{z},1}, \mathcal{D}_{\mathbf{z},2}) < \mathrm{BIC}(\mathcal{M}_1 \mid \mathcal{D}_{\mathbf{z},1}) + \mathrm{BIC}(\mathcal{M}_2 \mid \mathcal{D}_{\mathbf{z},2}) \tag{4.44}$$

or, equivalently,

$$-2 \log L + k \log n < -2 \log(L_1 L_2) + k_1 \log n_1 + k_2 \log n_2. \tag{4.45}$$

Informally, merging two models into one is beneficial if the joint model can explain the data equally well (i.e., $L \approx L_1 L_2$), while requiring only a single set of parameters.

If more than two trajectories are considered, one has to evaluate all possible assignments of these trajectories to models and select the assignment with the highest posterior. As this quickly becomes intractable due to the combinatorial explosion, we use an approximation and consider the trajectories sequentially and in the order in which the robot observes them. We check whether merging the new trajectory with one of the existing models leads to a higher posterior compared to adding a new model for that trajectory to the set of previously encountered models. This algorithm is summarized in Algorithm 2.

After having identified a set of models as prior information, we can exploit this knowledge to make better predictions when observing a previously unseen

Algorithm 2. Sequential clustering of kinematic trajectories

Input: ordered set of n trajectories
Output: ordered set of clustered models $\mathbb{M} = \{\mathcal{M}_1, \ldots, \mathcal{M}_m\}$ and corresponding trajectories $\mathbb{D} = \{\mathcal{D}_{\mathbf{z},1}, \ldots, \mathcal{D}_{\mathbf{z},m}\}$

Initially, $\mathbb{M} := \emptyset$ (and thus, $m = 0$);
for *each newly observed trajectory* $\mathcal{D}_{\mathbf{z},new}$ **do**
  `/* Either initialize a new model for the new trajectory.. */`
  Estimate a model $\mathcal{M}_{new}$ for $\mathcal{D}_{\mathbf{z},new}$;
  $\mathbb{M}_{best} := \mathbb{M} \cup \{\mathcal{M}_{new}\}$;
  $\mathbb{D}_{best} := \mathbb{D} \cup \{\mathcal{D}_{\mathbf{z},new}\}$;
  for $j \in \{1, \ldots, m\}$ **do**
    `/* ..or assign the new trajectory to one of the existing models */`
    Estimate a joint model $\mathcal{M}_{j+new}$ from $\mathcal{D}_{\mathbf{z},j} \cup \mathcal{D}_{\mathbf{z},new}$;
    if $p(\mathcal{M}_1, \ldots, \mathcal{M}_{j+new}, \ldots, \mathcal{M}_m) > p(\mathbb{M}_{best})$ **then**
      $\mathbb{M}_{best} := \{\mathcal{M}_1, \ldots, \mathcal{M}_{j+new}, \ldots, \mathcal{M}_m\}$;
      $\mathbb{D}_{best} := \{\mathcal{D}_{\mathbf{z},1}, \ldots, \mathcal{D}_{\mathbf{z},j} \cup \mathcal{D}_{\mathbf{z},new}, \ldots, \mathcal{D}_{\mathbf{z},m}\}$;
    end
  end
  $\mathbb{M} := \mathbb{M}_{best}$;
  $\mathbb{D} := \mathbb{D}_{best}$;
end
Return the set of clustered models $\mathbb{M}$ and corresponding trajectories $\mathbb{D}$;

articulated object. Consider the situation in which a partial trajectory of a new object has been observed. To exploit the prior information, we proceed exactly as before. We compute and compare the posteriors according to Eq. (4.45), i.e., we add the newly observed data points as a new model or merge them into one of the w previously identified models. At each time step, we check whether

$$p(\mathcal{M}_{new}, \mathcal{M}_1, \ldots, \mathcal{M}_w) < \max_{j=1,\ldots,w} p(\mathcal{M}_1, \ldots, \mathcal{M}_{j+new}, \ldots, \mathcal{M}_w). \quad (4.46)$$

If the newly observed data is merged with an existing model, the parameter vector is estimated from a much larger, combined dataset $\mathcal{D}_{\mathbf{z},j} \cup \mathcal{D}_{\mathbf{z},new}$ instead of $\mathcal{D}_{\mathbf{z},new}$ which may lead to a better estimation. Note that this step is carried out after each observation of the new sequence. Thus, if the currently manipulated object ceases to be explained by the known models, the method instantaneously creates a new model. After successful object manipulation, this model serves as additional prior information for the future.

As an concrete example, imagine that the robot explores cabinet doors in a kitchen scenario. Initially, the robot doesn't have any prior models. When it opens the first door, say a left-opening door, the robot observes the resulting

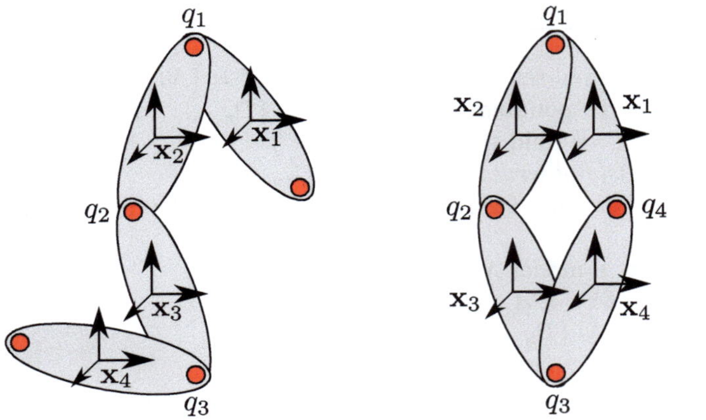

(a) open kinematic chain (b) closed kinematic chain

Fig. 4.4 Example of an open and a closed kinematic chain. (a) The open chain has three DOFs. (b) The closed chain has also only a single DOF.

opening trajectory and estimates a model for it. After this, the robot interacts with a right-opening door and observes its opening trajectory. Now, the robot has the option to either learn a new model for the right-opening door, or to merge the trajectories of both doors and learn a single, combined model. In this case, the combined model will have a very low data likelihood as the left- and right-opening doors are highly distinct. Therefore, the robot will prefer two separate models. When the robot continues its exploration of the kitchen, it might at some point observe the opening trajectory of another left-opening cabinet door. Again, the robot estimates a separate model but also tries to merge the trajectory into the already existing models. As the trajectories of the two left-opening doors are very similar (given that they have a similar radius and rotation axis), the combined model will explain both trajectories well (high data likelihood) while it requires only a single set of parameters. Correspondingly, the posterior probability of the combined model will be higher than the posterior probability of the separate models. Thus, the robot will decide to merge the new trajectory into the existing model for left-opening doors. As a result, the robot might interact with many cabinet doors, which it clusters into two distinct models: one for two left-opening doors and one for the right-opening door.

Closed Kinematic Chains

Although most articulated objects have the connectivity of kinematic trees, there are also mechanisms containing closed kinematic chains (or kinematic loops) (Featherstone and Orin, 2008). An intuitive example of a closed-loop system is given in Figure 4.4. The object in Figure 4.4a is a simple serial mechanism consisting of four parts and three revolute joints. When the two ends of this chain are linked, one obtains a closed kinematic chain as depicted in Figure 4.4b. Closed kinematic chains also appear when two open kinematic structures are combined. For example, consider a robot that opens a door with its manipulator. While both the robot and the door can be represented individually as kinematic trees using our approach, the combined system of the robot, the door, and the floor creates a kinematic loop. A similar mechanical system appears in case of a humanoid robot that has multiple contact points, for example, when standing on both feet, or a robot that manipulates an object with two arms (Sentis et al., 2010). To describe such closed-loop systems, we need to extend our approach.

Remember that we established in Section 4.1.3, that the best graph is the one that maximizes the posterior probability. Finding this graph was easy for kinematic trees because we could use the minimum spanning tree algorithm. However, finding the best kinematic graph is a more difficult problem. This results from the fact that the links are no longer independent of each other. In contrast to kinematic trees, the chained up predictions of the relative transformations between the object parts of a closed kinematic chain will, in general, not lead to a globally consistent prediction.

This problem, however, is closely related to loop-closing in the graph-based formulation of the simultaneous localization and mapping (SLAM) problem (Lu and Milios, 1997; Dellaert, 2005; Frese, 2006; Grisetti et al., 2009). For this type of problem, closed-form solutions exist only for very simple cases. A popular solution for the general case are iterative optimization approaches to deal with the underlying nonlinear least squares problem.

To obtain a consistent pose estimation for the whole graph, we use a pose graph optimization engine based on the Levenberg-Marquardt (LM) algorithm and originally designed to solve SLAM problems Grisetti et al. (2010). To generate the input graph, we proceed as follows. We add a vertex for each object part representing its initial pose $\hat{\mathbf{x}}'_1, \dots, \hat{\mathbf{x}}'_n$, that we estimate for an (arbitrary) spanning tree of the graph. Then, for each link model $\mathcal{M}_{ij}$ in our graph G, we add an edge that constrains the relative transformation between $\hat{\mathbf{x}}'_i$ and $\hat{\mathbf{x}}'_j$ to the expected transformation $\hat{\boldsymbol{\Delta}}_{ij}$ (in SLAM, this corresponds to an observation). The optimization procedure generates a set of corrected poses $\hat{\mathbf{x}}_1, \dots, \hat{\mathbf{x}}_n$ that it is the best prediction in terms of the squared error.

For the pose observations $\mathbf{y}_i$, we assume Gaussian noise with zero mean and covariance $\Sigma_{\mathbf{y}}$, i.e.,

$$\mathbf{y}_i = \mathbf{x}_i + \boldsymbol{\epsilon}, \tag{4.47}$$

$$\boldsymbol{\epsilon} \sim \mathcal{N}(0, \Sigma_{\mathbf{y}}). \tag{4.48}$$

The data likelihood of a single object part being observed at pose $\mathbf{y}$ while being expected at pose $\hat{\mathbf{x}}$ (and given the kinematic graph G and a configuration $\mathbf{q}$) then becomes

$$p(\mathbf{y}_i \mid G, \mathbf{q}) \propto \exp\left(-\frac{1}{2}(\hat{\mathbf{x}}_i \ominus \mathbf{y}_i)^T \Sigma_{\mathbf{y}}^{-1}(\hat{\mathbf{x}}_i \ominus \mathbf{y}_i)\right). \tag{4.49}$$

Using this, the *global* data likelihood of an articulated object in a particular configuration can be computed as the product over the likelihoods of all individual object parts, i.e.,

$$p(\mathbf{y}_1, \dots, \mathbf{y}_p \mid G, \mathbf{q}) = \prod_{i \in 1, \dots, p} p(\mathbf{y}_i \mid G, \mathbf{q}). \tag{4.50}$$

As the configuration $\mathbf{q}$ of the articulated object is latent and thus not known, we need to integrate over all possible configurations, i.e.,

$$p(\mathbf{y}_1, \dots, \mathbf{y}_p \mid G) = \int p(\mathbf{y}_1, \dots, \mathbf{y}_p \mid G, \mathbf{q}) p(\mathbf{q} \mid G)\, d\mathbf{q}. \tag{4.51}$$

Similar to Eq. (4.26), we approximate this integral by evaluating it only at the most likely configuration $\hat{\mathbf{q}}$ of the articulated object. We again assume that the configurations $\mathbf{q}$ are uniformly distributed, i.e., $p(\mathbf{q} \mid G) \approx n^{-D}$, where n is the number of pose observations and D are the DOFs of the articulated object. The data likelihood for a pose observation $\mathbf{y}_1, \dots, \mathbf{y}_p$ becomes

$$p(\mathbf{y}_1, \dots, \mathbf{y}_p \mid G) \approx n^{-D} p(\mathbf{y}_1, \dots, \mathbf{y}_p \mid G, \hat{\mathbf{q}}). \tag{4.52}$$

The data likelihood of an observation sequence $\mathcal{D}_{\mathbf{y}} = (\mathbf{y}_{1:p}^1, \dots, \mathbf{y}_{1:p}^n)$ of a whole articulated object is

$$p(\mathcal{D}_{\mathbf{y}} \mid G) \approx \prod_{i \in 1, \dots, n} n^{-D} p(\mathbf{y}_1^i, \dots, \mathbf{y}_p^i \mid G, \hat{\mathbf{q}}^i) \tag{4.53}$$

$$= n^{-nD} \prod_{i \in 1, \dots, n} p(\mathbf{y}_1^i, \dots, \mathbf{y}_p^i \mid G, \hat{\mathbf{q}}^i). \tag{4.54}$$

This data likelihood can now be used to compare alternative structures and to select the best one. Note that in principle, all possible graphs need to be evaluated – which is super-exponential in the number of object parts and polynomial in the number of template models, i.e., lies in $O\left((1+m)^{p^2}\right)$ for m candidate models and p object parts. In contrast, finding the exact solution in the case of kinematic trees has a polynomial complexity of $O(mp^2)$. Obviously, the massive set of possible graph structures can only be fully evaluated for small articulated objects and few template models.

In the absence of an efficient, exact solution, we propose an efficient approximation that is able to find the locally best graph from an initial guess using a randomized search strategy in polynomial time. The idea is that given the spanning tree as an initial solution, we iteratively evaluate the graphs in the neighborhood of the current best structure, i.e., graphs whose topology is similar to the current one, for example, by adding or removing one edge at a time. As we will see in the experimental section, this heuristic is able to find the optimal (or near-optimal) graph structure in most of the cases. Additionally, we can guarantee that this randomized search strategy never gets worse than the initial solution, which is in our case the spanning tree.

Finding the Degrees of Freedom

The current configuration of the whole articulated object is given by the stacked vector of all the individual configurations of its articulated links, i.e.,

$$\mathbf{q}_{\text{links}} = \begin{pmatrix} \mathbf{q}_{i_1 j_1} \\ \mathbf{q}_{i_2 j_2} \\ \vdots \\ \mathbf{q}_{i_m j_m} \end{pmatrix}, \tag{4.55}$$

where $\{(i_1 j_1), (i_2 j_2), \ldots, (i_m j_m)\} = E_G$. The question now is, whether the articulated object actually has as many DOFs as the sum of DOFs of its individual links might suggest. Clearly, in the case that the articulated object is a kinematic tree, the DOFs D_{object} of the articulated object directly equals the sum over the DOFs of its links $D_{\text{links}} = \sum_{(ij) \in E_G} d_{ij}$ as all of its links can be actuated independently of each other. However, for articulated objects containing loops, finding the DOFs of an articulated object is not trivial.

For an example, consider the object in Figure 4.4a which consists of three object parts and a total of three DOFs. In contrast, the object in Figure 4.4b consists of four object parts, connected by four revolute links in the form of a loop. Each of the four links has a single DOF and, therefore, the stacked configuration vector defining the configuration of all links is given by $\mathbf{q}_{\text{links}} = (q_1, q_2, q_3, q_4) \in \mathbb{R}^4$. Yet, the overall system has only a single DOF: when the first joint is brought into a particular configuration, the other joints are fixed as well, as a result of the loop closure. This means that the object configuration $q_{\text{object}} \in \mathbb{R}$ has only a single dimension and thus the object configuration space is a one-dimensional manifold embedded in the four-dimensional link configuration space.

Finding a mapping between the high-dimensional link configuration space $\mathbb{R}^{D_{\text{links}}}$ and a lower-dimensional object configuration space $\mathbb{R}^{D_{\text{object}}}$ can, for example, be obtained with PCA for linear manifolds or LLE for nonlinear manifolds. In the case of PCA, this correspond to finding a projection matrix $P \in \mathbb{R}^{D_{\text{object}} \times D_{\text{links}}}$ that describes the mapping

$$\mathbf{q}_{\text{object}} = P \mathbf{q}_{\text{links}}. \tag{4.56}$$

Recall from Eq. (4.54) that the DOFs has a strong influence on the data likelihood of a configuration because a higher dimensional configuration space results in a lower likelihood for a single configuration. As a result, a model with fewer DOFs is preferred over a model with more DOFs. At the same time, the dimension reduction of the configuration space adds additional parameters which need to be considered during model selection. For example, a linear projection requires a projection matrix P, and thus $k_{\text{PCA}} = D_{\text{object}} \cdot D_{\text{links}}$ additional parameters need (for the elements of the projection matrix) to be considered. These additional parameters, however, quickly pay off when the number of pose observations n increases. While the BIC penalty for additional model parameters grows with $\log n$ by the regularizer, the DOFs D are penalized linearly with $n \log n$ in the data likelihood term of the BIC score – see Eq. (4.54) and Eq. (4.34).

Informally speaking, if a kinematic graph with fewer DOFs explains the data equally well, it will have a slightly higher data likelihood, and thus, it will be favored in the structure selection step. In the experimental section, we will see that a robot can use this to accurately and robustly estimate the DOFs of various articulated objects.

4.3 Perception and Control of Articulated Objects

For estimating the kinematic model of an articulated object, our approach needs a sequence of pose observations $\mathcal{D}_{\mathbf{y}} = \langle \mathbf{y}^1_{1:p}, \ldots, \mathbf{y}^n_{1:p} \rangle$. For our experiments, we used two different sources for acquiring these pose observations. We used (1) visual markers to observe the object pose passively, and (2) the proprioception of the robot to observe the trajectory of its end effector while interacting with an object. Complementary to this, we will present in Chapter 5 an approach to marker-less pose perception using a stereo camera.

Marker-Based Perception

We used three different marker-based systems for observing the pose of an articulated object, each with different noise and outlier characteristics: a motion capture studio with low noise and no outliers, ARToolkit markers with relatively high noise and frequent outliers, and OpenCV's checkerboard detector with moderate noise and occasional outliers.

Motion capture studios typically use several high-speed cameras installed on a rig along the ceiling and markers attached on the individual parts of the articulated object, and provide highly accurate and virtually noise free pose estimates. The nominal noise of our PhaseSpace system is $\sigma_{\mathbf{y},\text{pos}} < 0.005\,\text{m}$ and $\sigma_{\mathbf{y},\text{orient}} < 1°$.

Additionally, we used the passive marker-based system ARToolkit for registering the 3D pose of objects by Fiala (2005). This system has the advantage that it requires only a single camera and can be used without any further infrastructure. The ARToolkit markers consist of a black rectangle and an error-correcting code imprinted on a 6x6-grid inside the rectangle for distinguishing the individual markers. We found that the observation noise of this system strongly depends on the distance and the angle of the marker to the camera. With a marker side length of 0.08 m and at a distance of 2 m from the camera, we typically obtain noise values of $\sigma_{\mathbf{y},\text{pos}} = 0.05\,\text{m}$ and $\sigma_{\mathbf{y},\text{orient}} = 15°$.

In contrast to this, OpenCV's checkerboard detector provides a much higher pose accuracy. The detector searches the camera images for strong black and white corners at sub-pixel accuracy (Bradski and Kaehler, 2008). With this system, we typically obtain measurement noise around $\sigma_{\mathbf{y},\text{pos}} = 0.005\,\text{m}$ and

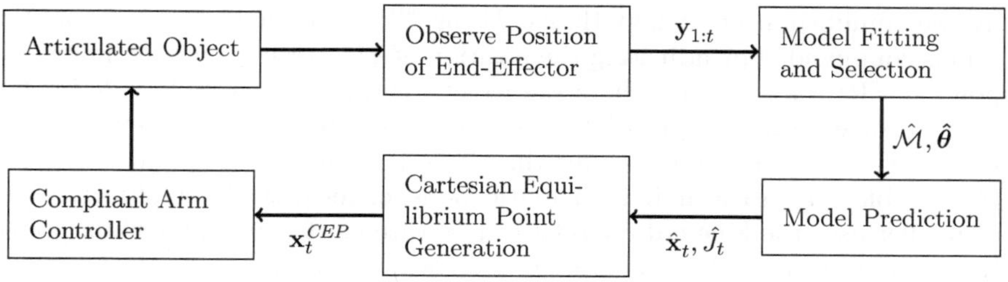

Fig. 4.5 Overall control structure. The robot iteratively estimates the kinematic model of the articulated object from the perceived trajectory of its end effector and evaluates it to generate the next Cartesian equilibrium point.

$\sigma_{\mathbf{y},\text{orient}} = 5°$ with marker sizes of 0.08 m side length at 2 m distance from the camera. One can distinguish different markers with this system by using checkerboards with varying numbers of rows and columns.

Real-Time Model Estimation and Manipulator Control

Next to visual observation of articulated objects, a mobile manipulation robot can also estimate the kinematic model while it physically interacts with an articulated object. When the robot establishes firm contact with the handle of a cabinet door, the position of its end effector directly corresponds to the position of the door handle. By evaluating its joint encoders, the robot can compute the pose of its gripper through its forward model. As a result, the robot can both sense the position of the handle as well as control it by moving the manipulator.

The approach described in this section was developed in collaboration with Jain and Kemp from the Healthcare Robotics Lab at Georgia Tech. The robot that we use for this research is a statically stable mobile manipulator named Cody. It consists of two arms from MEKA Robotics and an omni-directional mobile base from Segway. As end effector, it uses a hook inspired by prosthetic hooks and human fingers, which is described in more detail in the recent work of Jain and Kemp (2009a). Furthermore, we used a PR2 robot from Willow Garage for additional experiments, equipped with a standard two-finger parallel-yaw gripper.

Figure 4.5 shows a block diagram of our controller. The robot observes the pose of its end effector in Cartesian space, denoted by $\mathbf{y} \in \mathbb{R}^3$. While operating the object, the robot records the trajectory $\mathbf{y}_{1:t}$ over time as a sequence of poses. From this partial trajectory, it continuously estimates the kinematic model of the articulated object, that the robot uses in turn to predict the continuation of the trajectory.

To actually operate an articulated object, we use equilibrium point control (EPC) (Jain and Kemp, 2010) which is a form of impedance control inspired by the equilibrium point hypothesis. Using EPC, the motion of the robot's arm is commanded by adjusting the position of a Cartesian-space equilibrium point (CEP) that denotes where the robot's end effector would settle in the absence of externally applied forces other than gravity. We developed a trajectory controller that updates the Cartesian equilibrium point based on the Jacobian of the estimated kinematic model of the articulated object. This controller uses the kinematic model to generate Cartesian equilibrium point trajectories in a fixed world frame, attached to the initial location of the handle. At each time step t, the controller computes a new equilibrium point $\mathbf{x}_t^{CEP}$ as

$$\mathbf{x}_t^{CEP} = \mathbf{x}_{t-1}^{CEP} + \mathbf{v}_t^{mechanism} + \mathbf{v}_t^{hook}, \tag{4.57}$$

where $\mathbf{v}_t^{mechanism}$ is a vector intended to operate the mechanism and $\mathbf{v}_t^{hook}$ is a vector intended to keep the hook from slipping off the handle. The controller computes

$$\mathbf{v}_t^{mechanism} = s\frac{\hat{J}_t}{\|\hat{J}_t\|} \tag{4.58}$$

as a vector of length $s = 0.01\,\text{m}$ along the Jacobian of the learned kinematic function of the mechanism, i.e.,

$$\hat{J}_t = \nabla f_{\hat{\mathcal{M}},\hat{\boldsymbol{\theta}}}(\mathbf{q})\big|_{\mathbf{q}=\mathbf{q}^t}. \tag{4.59}$$

For $\mathbf{v}_t^{hook}$, we use a proportional controller that tries to maintain a force of 5 N between the hook and the handle in a direction perpendicular to $\hat{J}_t$. This controller uses the force measured by the wrist force-torque sensor of the robot. We refer the reader to the work of Jain and Kemp (2009b) for details about the implementation of equilibrium point control and how it can be used to coordinate the motion of a mobile base and a compliant arm (Jain and Kemp, 2010).

The positional accuracy of the manipulator itself is very high, i.e., $\sigma_{\mathbf{y},\text{pos}} \ll 0.01\,\text{m}$. However, by using a hook as the end effector, the robot cannot sense the orientation of the handle. As the manipulator is mounted on a mobile base, the robot can move around and thus the positional accuracy of the sensed position of the hook in a global coordinate system (and thus including localization errors of the base) reduces to about $\sigma_{\mathbf{y},\text{pos}} \approx 0.05\,\text{m}$.

4.4 Experiments

In this section, we present the results of a thorough evaluation of all aspects of our framework. First, we demonstrate that accurate and robust estimates of the kinematic models can be obtained using artificial markers. Second, we show that our approach also works on data acquired with two different mobile manipulation robots operating various pieces of furniture in domestic

Table 4.2 Evaluation of model prediction errors of the articulation models learned of a microwave oven, an office cabinet, and a garage door.

Articulated object		Rigid model	Prismatic model	Revolute model	GP model
Microwave	pos. error =	0.3086 m	0.1048 m	**0.0003 m**	0.0020 m
($\sigma_{\mathbf{z},\text{pos.}} = 0.002\,\text{m}$,	orient. error =	37.40°	32.31°	**0.15°**	0.16°
$\sigma_{\mathbf{z},\text{orient.}} = 2.0°$)	$\gamma =$	0.891	0.816	**0.000**	0.000
Drawer	pos. error =	0.0822 m	**0.0016 m**	0.0018 m	0.0017 m
($\sigma_{\mathbf{z},\text{pos.}} = 0.002\,\text{m}$,	orient. error =	2.06°	**1.36°**	1.60°	1.09°
$\sigma_{\mathbf{z},\text{orient.}} = 2.0°$)	$\gamma =$	0.887	**0.000**	0.003	0.000
Garage door	pos. error =	1.0887 m	0.3856 m	0.4713 m	**0.0450 m**
($\sigma_{\mathbf{z},\text{pos.}} = 0.050\,\text{m}$,	orient. error =	14.92°	10.79°	10.34°	**0.93°**
$\sigma_{\mathbf{z},\text{orient.}} = 5.0°$)	$\gamma =$	0.719	0.238	0.418	**0.021**

environments and thus is applicable to a wide range of manipulation tasks for mobile service robots. Third, we present a detailed analysis of our method on synthetic data. In particular, we study the robustness of model estimation and selection against noise and outliers, the convergence behavior, and the computational complexity as a function of the number of training samples.

4.4.1 Model Estimation and Model Selection

For our first experiments, we observed the poses of three typical objects in domestic environments: the door of a microwave oven, the drawers of an office cabinet, and a garage door. The goal of these experiments is to show that our approach both robustly and accurately estimates link models, as well as the correct kinematic structure of the whole object. In addition, we demonstrate that the range of the configuration space can be obtained during model estimation.

We tracked the motion of the microwave oven and the cabinet using a motion capture studio and the garage door using checkerboard markers. For each object, we recorded 200 data samples while we manually articulated each object. To evaluate our system, we used 10-fold cross-validation. For each of the 10 runs, we sampled $n = 20$ observations that we use for fitting the model parameters. We used the remaining observations for measuring the prediction accuracy of the fitted model.

Model Fitting

The quantitative results of model fitting and model selection are given in Table 4.2. As can be seen from this table, the revolute model is well suited for predicting the opening movement of the microwave door (error below 0.0001 m) while the prismatic model predicts very accurately the motion of

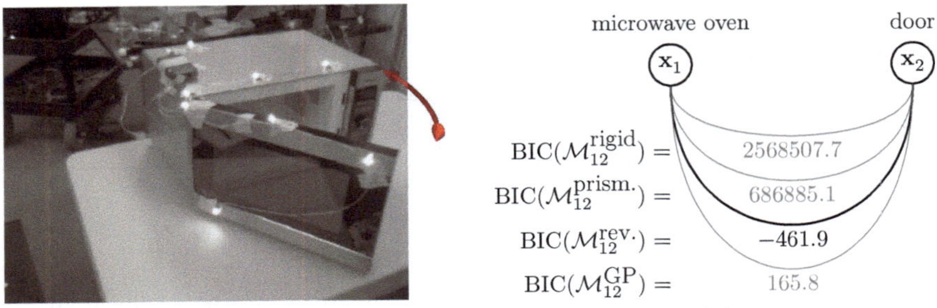

(a) (b)

Fig. 4.6 Visualization of the kinematic model learned for the door of a microwave oven. (a) Visualization of the configuration range. (b) Kinematic graph. The numbers on the edges indicate the BIC score of the corresponding model candidate.

the drawer (error below 0.0016 m), which is the expected result. Note that the revolute model can also explain the motion of the drawer with an accuracy of 0.0017 m by estimating a revolute joint with a large radius. It should be noted that the flexible GP model provides roughly the same accuracy as the parametric models and is able to robustly predict the poses of both datasets (0.0020 m for the door and 0.0017 m for the drawer). In the case of the simulated garage door, however, all parametric models fail whereas the GP model provides accurate estimates. The reader might wonder now why the GP model alone does not suffice, as the GP model can represent many different types of kinematic models, including revolute and prismatic ones. However, the GP model has a variable complexity and is thus more prone to over-fitting in the presence of noise. In contrast, the specialized models have a smaller number of free parameters and are therefore more robust against noise and outliers. Furthermore, they require less observations to converge. We will investigate these properties in more detail in Section 4.4.4. These experiments illustrate that our system takes advantage of the expert-designed parametric models when appropriate while we have the flexibility to learn models for unforeseen mechanical constructions as well.

The learned kinematic models also provide the configuration range C of the articulated object. For visualization purposes, we can now sample configurations from this range and project them to object poses using the learned forward function. Figures 4.6, 4.7, and 4.8 illustrate the learned configuration range for the door of the microwave oven, the garage door, and the two drawers of the office cabinet, respectively.

Model and Structure Selection

After fitting the model candidates to the observed data, the next goal is to select the model that best explains the data, which corresponds to finding the model that maximizes the posterior probability (or minimizes the BIC score).

The right image in Figure 4.6 shows the resulting graph for the microwave oven dataset, with the BIC score indicated at each edge. As expected, the revolute model is selected, because it has the lowest BIC score. Correspondingly, the right image in Figure 4.7 shows the BIC scores for all edges for the garage door dataset, where the GP model gets selected.

A typical articulated object consisting of multiple parts is a cabinet with two drawers as depicted in Figure 4.8. In this experiment, we track the poses of the cabinet itself ($\mathbf{x}_1$) and its two drawers ($\mathbf{x}_2$ and $\mathbf{x}_3$). During the first 20 samples, we opened and closed only the lower drawer. Accordingly, a prismatic joint model $\mathcal{M}_{23}^{\text{prism.}}$ is selected (see top row of images in Figure 4.8). When also the upper drawer gets opened and closed, the rigid model $\mathcal{M}_{12}^{\text{rigid}}$ is replaced by a prismatic model $\mathcal{M}_{12}^{\text{prism.}}$, and $\mathcal{M}_{23}^{\text{prism.}}$ is replaced by $\mathcal{M}_{13}^{\text{prism.}}$, resulting in the kinematic tree $E_G = \{(1,2),(1,3)\}$. Note that it is not required to articulate the drawers one after each other. This was done only for illustration purposes.

Multi-dimensional Configuration Spaces

To illustrate that our approach is also able to find models with higher-dimensional configuration spaces with $d > 1$, we let the robot monitor a table that was moved on the floor. The robot is equipped with a monocular camera tracking an ARToolkit marker attached to the table. In this experiment, the table was only moved but never turned, lifted, or tilted and, therefore, the observable configuration space of the table has two dimensions. Figure 4.9 shows four snapshots during learning. Initially, the table is perfectly explained as a rigid object in the room (a). Then, a prismatic joint model best explains the data since the table was moved in one direction only (b). After moving sideways, a 1-DOF Gaussian process model is selected that follows a simple curved trajectory (c). Finally, the full planar movement is explained by a 2-DOF Gaussian process model, that can model movements on two-dimensional surfaces (d).

Additional Examples

We ran similar experiments on a large set of different articulated objects that typically occur in domestic environments, including office cabinets, office doors, desk lamps, windows, kitchen cabinets, fridges, dishwashers, and garage doors. Four examples are given in Figure 4.10. For these experiments, we attached checkerboards of different sizes to all movable parts and used both a consumer-grade video camera and a low-cost laptop webcam to acquire the image data. Our software also visualizes the learned articulation models in 3D and back-projects them onto the image to allow for easy visual inspection. The detected poses of the checkerboards are visualized as red/green/blue coordinate axes systems, and selected links between them are indicated using a colored connection. The software also displays the

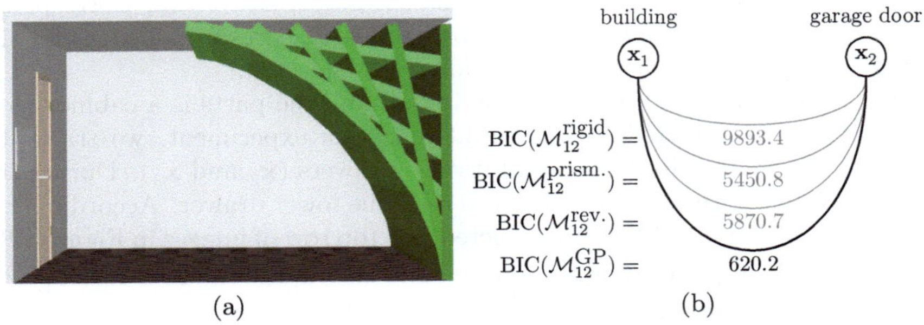

Fig. 4.7 Visualization of the kinematic model learned for a garage door. (a) Ten uniformly sampled configurations from the learned model. (b) Corresponding kinematic graph.

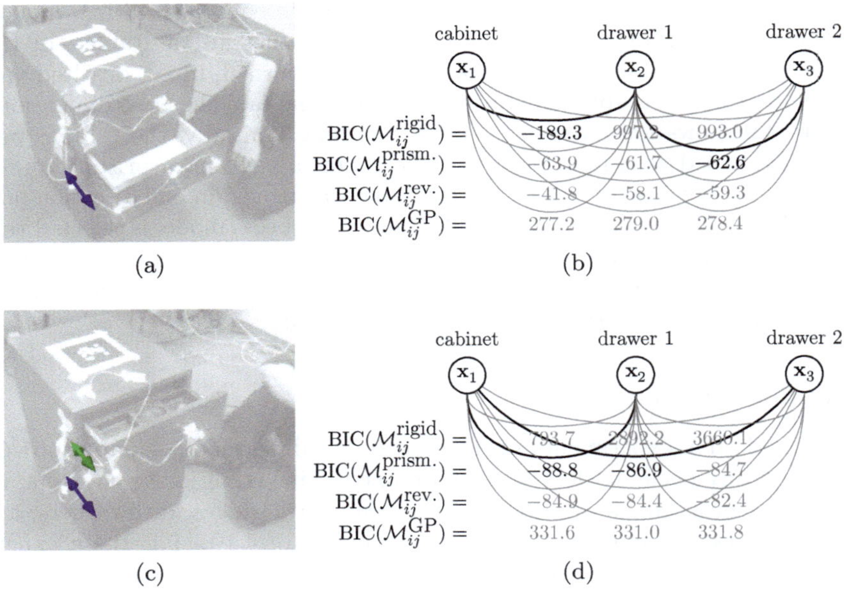

Fig. 4.8 Incrementally estimating a model of two drawers of a cabinet. (a)+(b) Initially, only the lower drawer is opened and closed. (c)+(d) Both drawers are opened and closed independently.

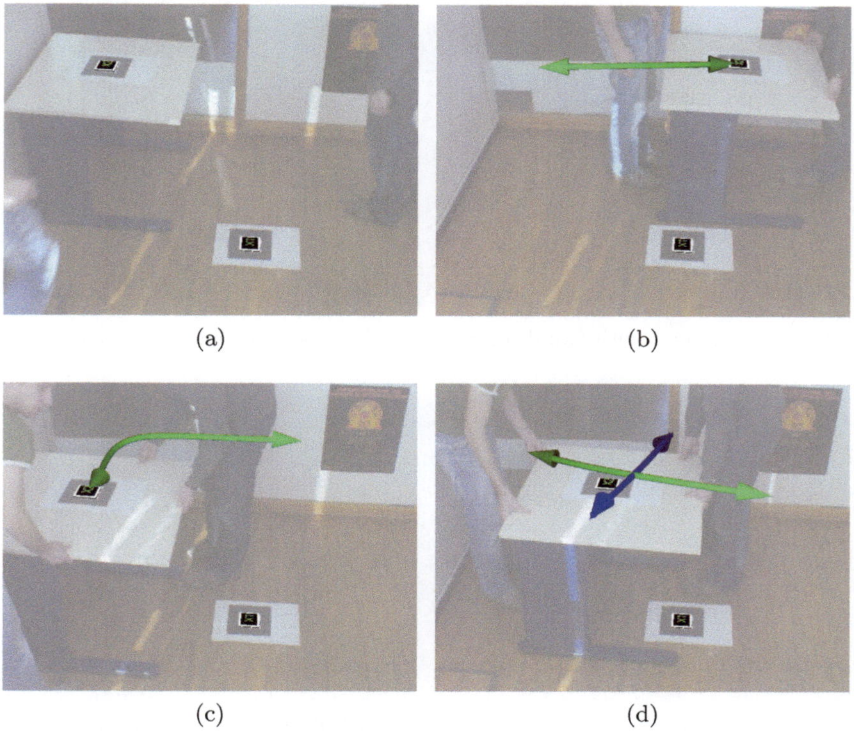

Fig. 4.9 Learning a model for a table moving on the floor. The arrows indicate the recovered manifold of the configuration space.

configuration range by generating poses in the estimated range. For revolute joints, it additionally indicates the rotation axis using a line and a surrounding circle.

From visual inspection of the objects in Figure 4.10, one can see how accurate the model estimation works in conjunction with marker-based tracking: the motion of the drawers of the cabinet is well matched, and the rotation axes of the door hinge and the door handle are estimated very close to their true position. The upper part of the garage door moves in a slider in the ceiling while the lower part is connected via a revolute joint. The resulting motion is clearly neither revolute nor prismatic, and consequently our approach selects the GP model. The desk lamp consists of two-bar links that keep the light housing always upright (or, loosely speaking, rigid in orientation), but moves its head along a circle. This link type can be well explained by the GP model. The existence of these objects shows the necessity to supply a domestic service robot with such a general, nonparametric model that can deal with a wide variety of different articulated objects. Yet, it is also clear that the majority of articulated objects in domestic environments will consist

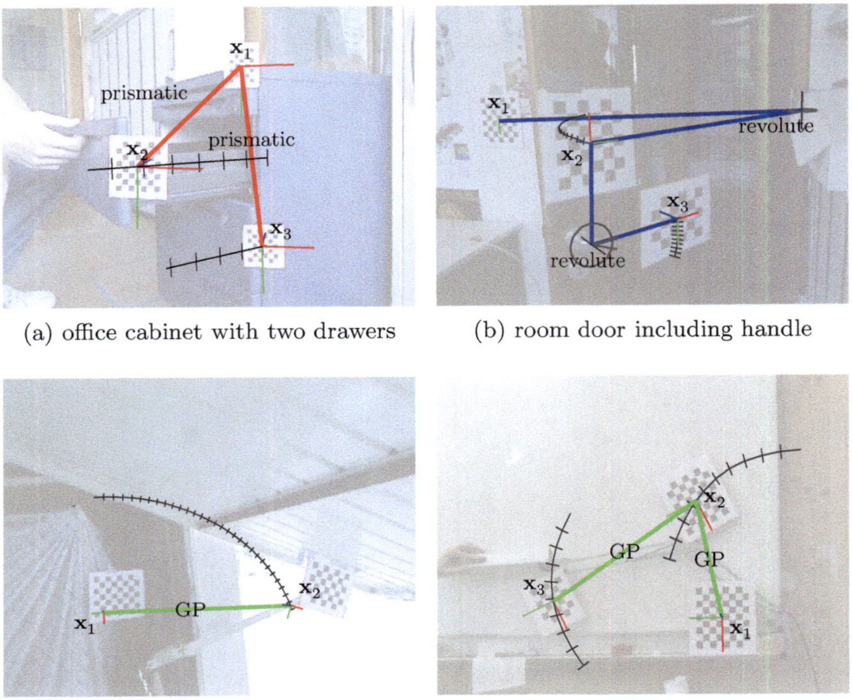

(a) office cabinet with two drawers

(b) room door including handle

(c) garage door

(d) desk lamp with two links

Fig. 4.10 Visualization of the learned articulation models for several further domestic objects.

of revolute and prismatic joints which can be more robustly estimated using parametric models. This motivates again the necessity to fit both parametric and nonparametric models to obtain the best performance.

Another interesting object is a car, as its doors and windows have both tree- and chain-like elements. In Figure 4.11, we observed the motion of the driver's door and window. After the first few observations, our approach estimated the structure to be rigid and links both the door and the window in parallel to the car body. After we opened the window to the half, our approach attached the driver's window to the door and selected a prismatic model. Surprisingly to us, when we opened the window further (and thus acquire more observations), our approach switched to a revolute model for the driver's window associated with a large radius ($r = 1.9\,\mathrm{m}$). By looking carefully at the data and the car, we can confirm that the window indeed moves on a circular path which is due to its curved window glass. Finally, after the driver closed the door, also a revolute model for the link between the car body and the door was selected.

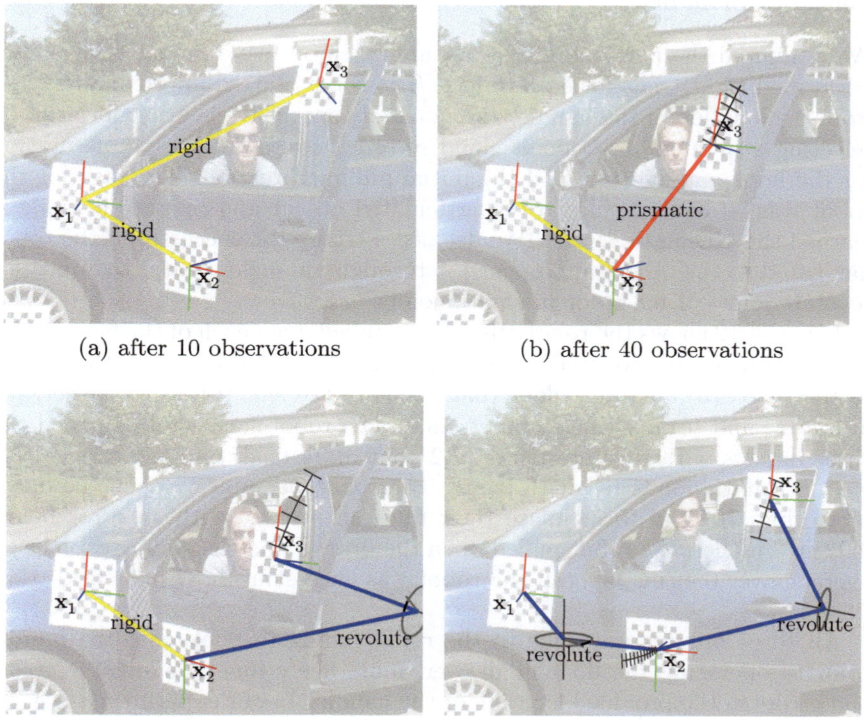

(a) after 10 observations (b) after 40 observations

(c) after 60 observations (d) after 140 observations

Fig. 4.11 Snapshots of the learning process when incrementally observing the motion of a car door and its window from camera images.

We conclude from these results, that our approach is able to estimate both the kinematic parameters and kinematic structures of several objects relevant for domestic service robots at high accuracy, i.e., the prediction error of the learned models is around 0.001 m and 1° for objects tracked in a motion capture studio, and around 0.003 m and 3° for checkerboard markers. At this accuracy, the learned models are well suited for robotic manipulation tasks.

4.4.2 Operating Articulated Objects with a Mobile Manipulator

In this section, we show that real robots can utilize our approach to learn the kinematic models of objects for active manipulation. Here, control of the arm was implemented in close collaboration with Jain and Kemp. The experiments were conducted on two different platforms, the robot "Cody" and "Marvin" robot (see Figure 1.2).

Task Performance

We evaluated the performance of our approach on five different objects and performed eight trials for each object. The robot started approximately 1 m from the location of the handle. We manually specified the grasp location by selecting a point in a 3D point cloud recorded by the robot, an orientation for the hook end effector, and the initial pulling direction. The task for the robot was to navigate up to the articulated object and operate it while it learned the kinematic model. We deemed a trial to be successful if the robot navigated to the object and opened it through an angle greater than 60° for revolute joints or 0.3 m for prismatic joints.

Figure 4.12 shows the robot after it has pulled open each of the five objects in one of the respective trials. The objects are: (a) a cabinet door that opens to the right, (b) a cabinet door that opens to the left, (c) a dishwasher, (d) a drawer, and (e) a sliding cabinet door. The robot successfully opened the three objects with revolute joints in 21 out of 24 trials and the two objects with prismatic joints in all 16 trials. The robot was able to open the doors more than 70° and to estimate their radii on average with an error below 0.02 m. Further, the robot pulled open the drawer and the sliding cabinet on average over 0.49 m. Overall the robot was successful in 37 out of 40 trials (92.5 %).

All three failures were due to the robot failing to hook onto the handle prior to operating the door or drawer, most likely due to odometry errors and errors in the provided location of the handle. In our experiments, we did not observe that the model learning caused any errors. In principle, however, the hook could slip off the handle if a wrong model had been estimated.

Model Fitting and Selection from End-Effector Trajectories

Figure 4.1 and Figure 4.13 show examples of the PR2 robot operating several articulated objects common to domestic environments, i.e., a fridge, a drawer, a dishwasher door, the tray of a dishwasher, and the valve of a heater. For these experiments, we did not use the feedback control loop as described in Section 4.3 but operated the manipulator manually. First, we recorded a set of trajectories by guiding the manipulator to operate various articulated objects. During execution, we played these trajectories back using a different implementation of equilibrium point control available on the PR2 platform and recorded the resulting end effector trajectories of the robot. We used these trajectories subsequently to learn the kinematic models. Finally, we visualized these models by superimposing them on images taken by a calibrated wide-angle camera mounted on the head of the robot. In our experiments, our approach always selected the correct model candidate. One can easily verify by visual inspection that our approach estimates the kinematic properties (such as the rotation axis or the prismatic axis) very accurately.

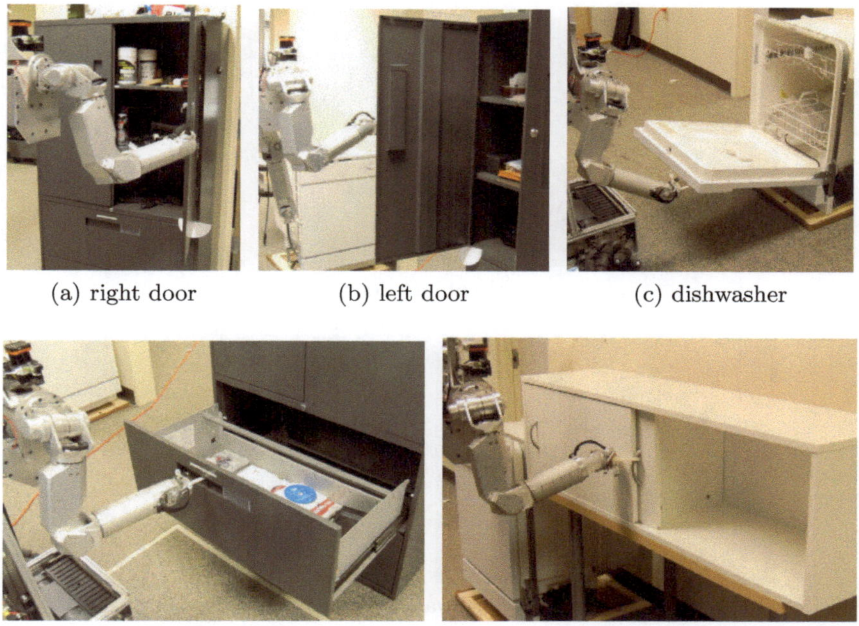

(a) right door (b) left door (c) dishwasher

(d) drawer (e) sliding door

Fig. 4.12 Images showing Cody at Georgia Tech operating the five different objects using the approach described in Section 4.3. Images courtesy of Jain and Kemp.

These experiments show that robots can successfully learn accurate kinematic models of articulated objects from end effector trajectories using our approach. With the PR2, we achieved an average predictive accuracy of the learned models below 0.002 m, which is more than sufficient for using our models for mobile manipulation tasks in domestic settings.

Improving Model Estimation based on Experience

In the experiments described in the previous section, we learned the kinematic models for the kitchen furniture independently of each other. By using the approach described in Section 4.2 on data from Cody, we verified that prior experience supports model learning. Figure 4.14 shows the result of this experiment. The colors indicate the prior models to which our approach assigned the observed trajectories. Our approach correctly recognized that the robot had operated five different objects and assigned the 37 different trajectories correctly to the corresponding models.

We measured the average prediction error with and without learning prior models (see Figure 4.15), using leave-one-out cross-validation and a randomized ordering of the trajectories. We found that the prior models reduce the

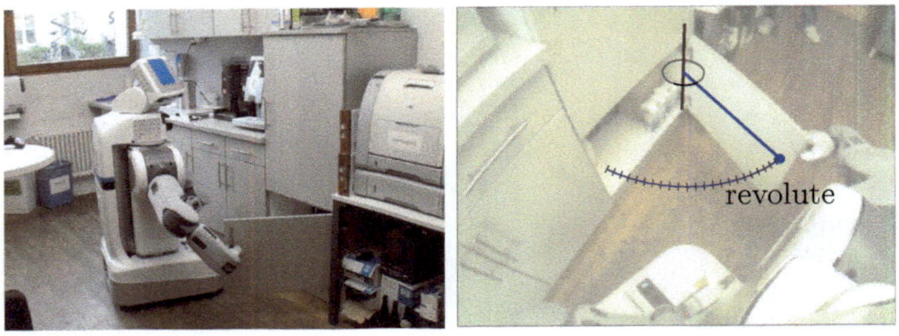

(a) cabinet door

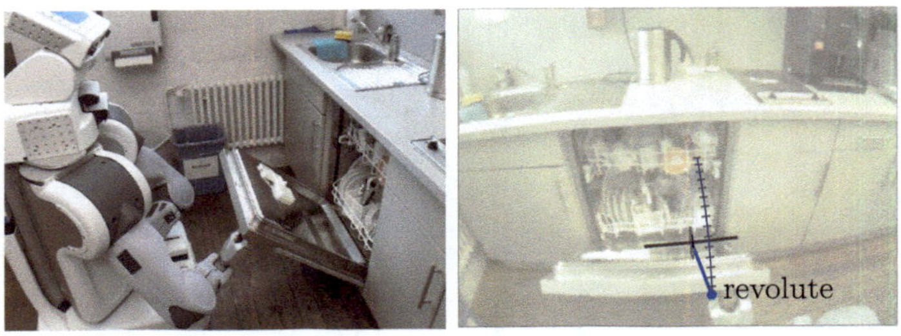

(b) dishwasher door

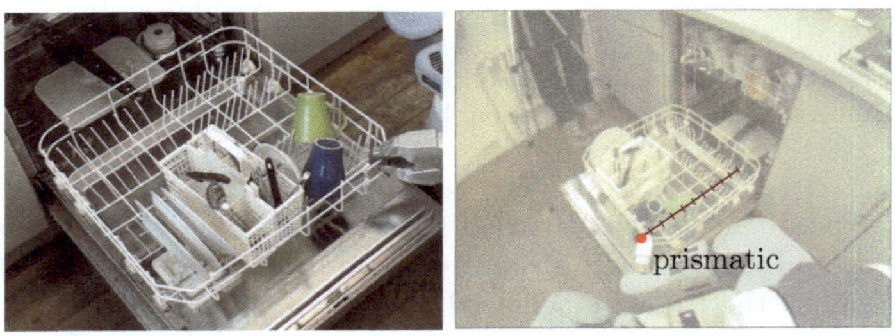

(c) dishwasher tray

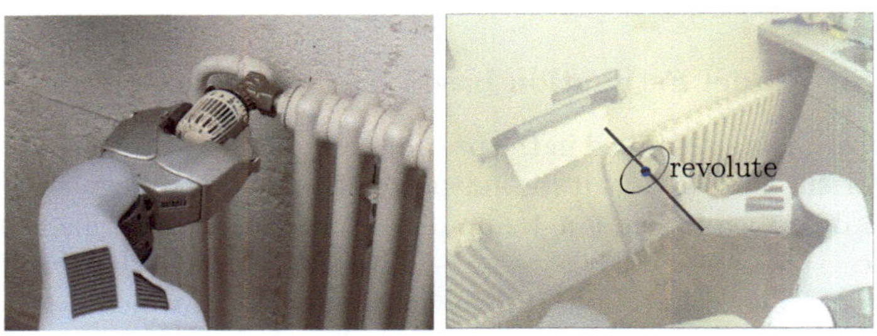

(d) valve of a heater

Fig. 4.13 A PR2 robot learns the kinematic models of different pieces of furniture while it operates them.

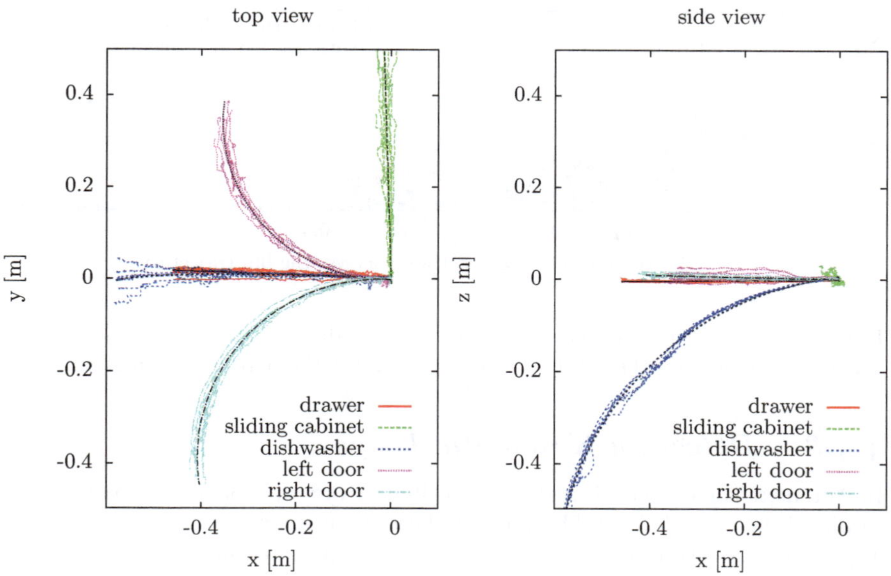

Fig. 4.14 Observed trajectories and the 5 recovered models when minimizing the overall BIC using our approach. Trajectories assigned to the same model are depicted in the same color.

prediction error considerably, especially if the new trajectory is only partially observed. When 30 % to 70 % of the new trajectory had been observed, the prediction error was reduced by a factor of three and more. As a result, the robot came up with a substantially more accurate model early and could utilize this knowledge to improve the control of its manipulator.

Throughout all experiments on Cody, we used a fixed noise term of $\sigma_{\mathbf{z},\text{pos}} = 0.05\,\text{m}$. This accounts for inaccuracies in the observation of the end effector position due to variations in the hooking position, and small errors in the kinematic forward model and robot base localization. We found in repeated experiments that in the range between $0.02\,\text{m} \leq \sigma_{\mathbf{z},\text{pos}} \leq 0.20\,\text{m}$, the results are similar to our previous results obtained with $\sigma_{\mathbf{z},\text{pos}} = 0.05\,\text{m}$. Only for significantly smaller values of $\sigma_{\mathbf{z},\text{pos}}$ more prior models are created, for example due to small variations of the grasping point and other inaccuracies. For much larger values, observations from different objects are clustered into a joint model. Thus, our results are insensitive to moderate variations in the observation noise $\sigma_{\mathbf{z},\text{pos}}$.

This experiment illustrates that our approach enables a mobile robot to learn from experience or exploit prior information when manipulating new objects. The experience increases the prediction accuracy by a factor of approximately three.

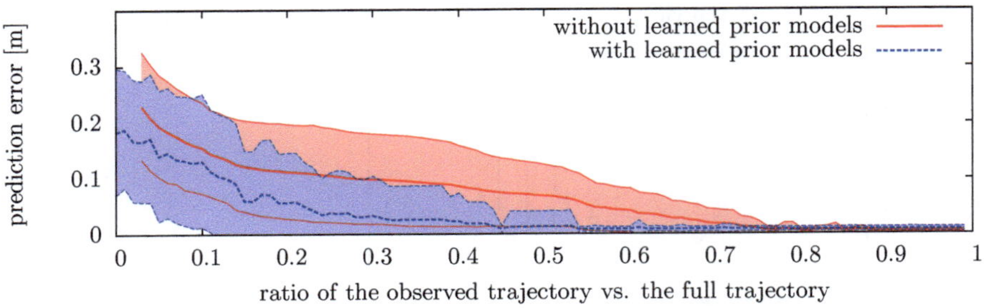

Fig. 4.15 Average prediction error (line) and standard deviation (shaded area) of the learned model on the full trajectory with and without prior information.

4.4.3 Detecting Kinematic Loops

In our next set of experiments, we evaluated our approach on objects containing kinematic loops. The goal of these experiments is to show that our approach can estimate correctly both the kinematic connectivity, as well as the DOFs.

For this purpose, we used the first four segments of a yardstick. This results in an open kinematic chain consisting of three revolute joints (see Figure 4.16a). This object has three DOFs, as all revolute joints are independent of each other. In a second experiment, we taped the fifth segment of the yardstick together with the first one which created a kinematic loop (see Figure 4.16b). The resulting object consists of four revolute joints each having a single DOF. The resulting object has effectively only a single DOF. We articulated the objects manually and recorded object pose datasets with $|\mathcal{D}_{\mathbf{y}}| = 200$ samples each using checkerboard markers.

Figures 4.16a–f visualize the learned kinematic model for the open and the closed kinematic model, respectively, while Figures 4.16g+h show the kinematic structure of the learned model. From this figure, it can be seen that our approach correctly recognized that the object with an open kinematic consists of three revolute links ($\mathcal{M}_{12}^{\text{rot.}}, \mathcal{M}_{23}^{\text{rot.}}, \mathcal{M}_{34}^{\text{rot.}}$), having three DOFs $\mathbf{q} = (q_1, q_2, q_3)$ in total. For the object with the kinematic loop, our approach selected four revolute links ($\mathcal{M}_{12}^{\text{rot.}}, \mathcal{M}_{23}^{\text{rot.}}, \mathcal{M}_{34}^{\text{rot.}}, \mathcal{M}_{14}^{\text{rot.}}$) and correctly inferred that the object only exhibits a single DOF $\mathbf{q} = (q_1)$.

We also analyzed the progression of model selection while the training data is incorporated. Figure 4.17a shows the DOFs of the learned kinematic model for the open kinematic chain. Note that we opened the yardstick segment by segment, therefore the DOFs increases step-wise from zero to three. Figure 4.17b shows the estimated DOFs for the closed kinematic chain: our approach correctly estimated the DOFs to one already after the first few observations.

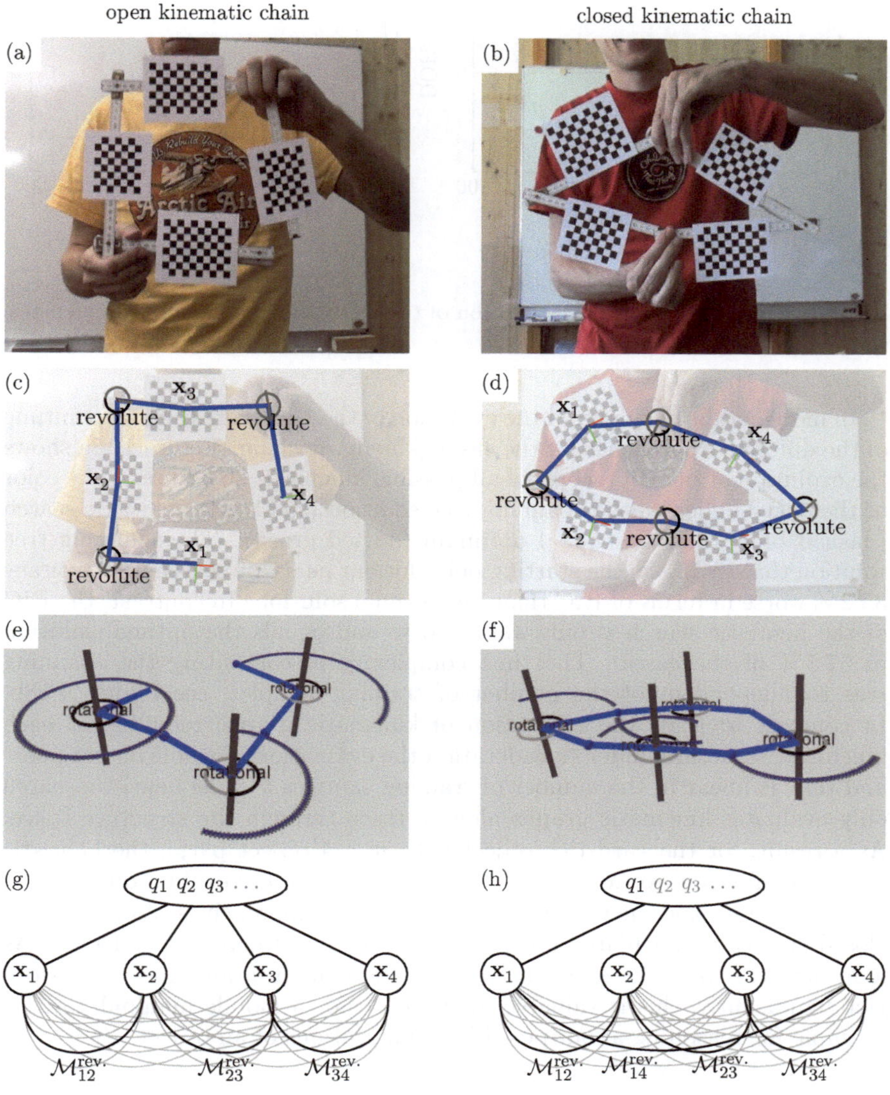

Fig. 4.16 Open kinematic chain with three DOFs (left column) and closed kinematic chain with only a single DOF (right column). (a)+(b) Images of the two objects. (c)–(f) Learned kinematic models from two different perspectives. (g)+(h) Learned graphical models, showing the connectivity and DOFs of the learned kinematic models. The selected kinematic model is visualized by bold edges, the estimated DOFs are given by the boldly type-set configuration variables.

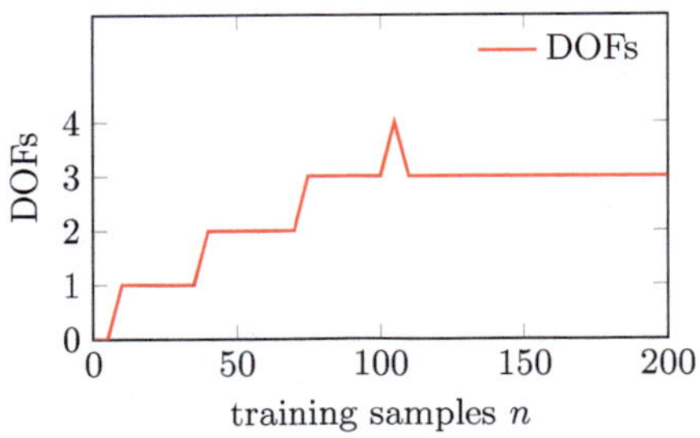

(a) open kinematic chain

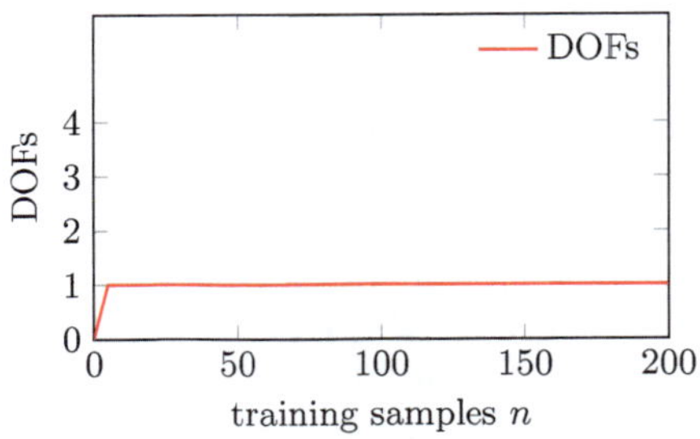

(b) closed kinematic chain

Fig. 4.17 Experiment on the estimation of the DOFs of open and closed kinematic chains.

In more detail, we analyzed the evolution of the BIC scores and the runtime of the different approaches for the closed kinematic chain. Figure 4.18a shows the evolution of the BIC scores of all possible kinematic structures. The color of the curve indicates the spanning tree solution (solid red), heuristic search (dashed blue), and the global optimum (dotted green). The spanning tree solution that we use as the starting point for our heuristic search is on average 35.2 % worse in terms of BIC than the optimal solution. In contrast, the BIC of the heuristic search is only 4.3 % worse and equals the optimal solution in 57.5 % of the cases. The time complexity of computing the spanning tree is independent of the number of training samples, see Figure 4.18b. In contrast to that, the evaluation of kinematic graphs requires for each kinematic structure under consideration the evaluation of whole object poses and thus is linear in the number of training samples n. The heuristic search only evaluates kinematic graphs along a trace through the structure space. As a result, for the yardstick object with $p = 4$ object parts, the heuristic search requires on average 82.6 % less time than the full evaluation.

With these experiments, we showed that our approach is able to detect closed chains in articulated objects and correctly estimates their DOFs. As loop closures (or reduced DOFs) reduce the configuration space of an object significantly, this is valuable information for a mobile manipulator, for example for reasoning about possible configurations of an object.

4.4.4 Robustness and Convergence Analysis

In this section, we report our results of a detailed robustness and convergence analysis of the candidate models proposed in this chapter. In particular, we analyzed the accuracy and the robustness against varied amounts of noise and outliers, and the convergence behavior with respect to the number of training

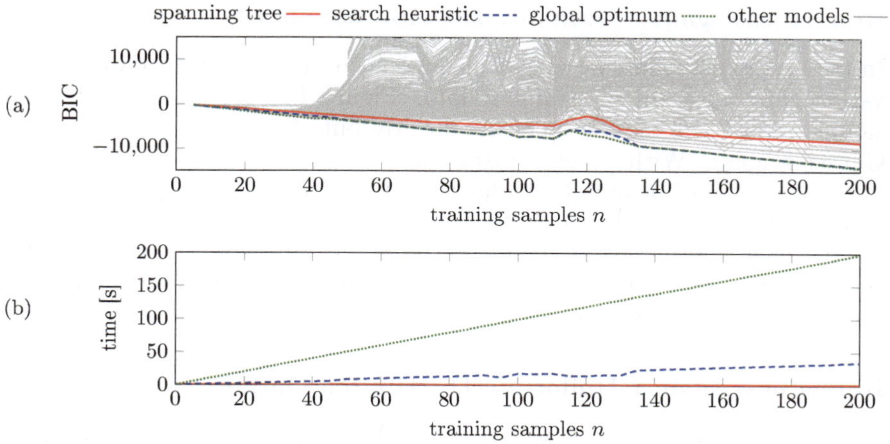

Fig. 4.18 Evaluation of our algorithm on closed kinematic chains. (a) BIC scores of all considered kinematic structures. (b) Computation times as a function of the number of training samples using different strategies.

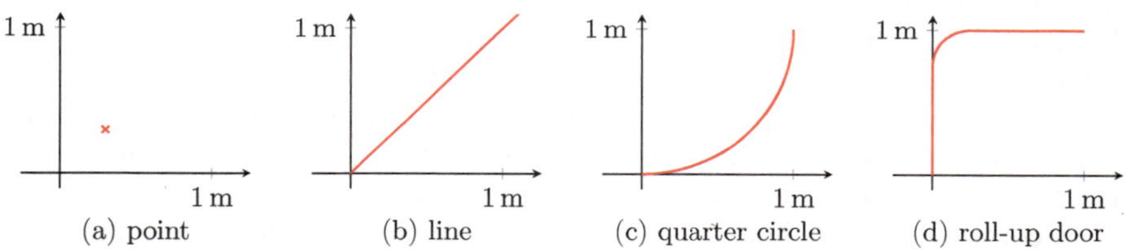

Fig. 4.19 The four ground truth models used for the evaluation of the estimators.

samples. The goal of these experiments was to show that our model estimators are robust against reasonable amounts of noise and other disturbances.

We conducted these experiments using synthetic datasets which allowed us to control the parameters for noise $\Sigma_{\mathbf{z}}$, outlier ratio γ, and the number of samples n. For each run, we generated an observation sequence from a known model $\mathcal{M}, \boldsymbol{\theta}$ with outlier ratio γ and noise level $\sigma_{\mathbf{z},\text{pos}}$. From that, we estimated the model's parameter vector $\hat{\boldsymbol{\theta}}$ using our approach. To measure the error of the estimated model, we generated an observation sequence of noise-free and outlier-free testing samples.

For each model, we chose a set of fixed ground truth models that we used for simulation. These models are depicted in Figure 4.19 and were used to sample observation sequences according to our observation model. All experiments reported in this section were repeated and averaged over 100 independent runs. If not stated otherwise, we sampled for each run a sequence of $n = 100$ observations.

Robustness against Normally-Distributed Noise

In the first experiment, we varied the amount of normally distributed observation noise $\sigma_{\mathbf{z},\text{pos}}$ in the training data. Figure 4.20 gives the result. The sample error of all models increases linearly but slowly with the amount of Gaussian noise. With an observation noise of $\sigma_{\mathbf{z},\text{pos}} = 1\,\text{m}$, the GP model yields a sample error of 0.34 m, the revolute model of 0.28 m, the prismatic model of 0.22 m, and the rigid model of 0.14 m. Thus, the rigid model is by far the most robust against Gaussian noise. We attribute this to the fact that the rigid model needs to estimate the fewest parameters and thus has the highest ratio of training samples to free parameters.

Robustness against Uniformly-Distributed Outliers

In a second experiment, we varied the outlier ratio $0 \leq \gamma \leq 1$. The result is given in Figure 4.21: the rigid, prismatic, and revolute model, i.e., the models using MLESAC estimators, are highly robust against outliers. Even when 50 % of the training data are real outliers, the sample error is only twice as high as when the training data is completely outlier-free. The rigid model is the most robust (its sample error stays below the observation noise level until $\gamma = 0.95$), followed by the prismatic model ($\gamma = 0.8$), and the revolute model ($\gamma = 0.65$). We attribute this effect to the model complexity, i.e., models with more parameters require more samples for estimating the parameters. We expect that more iterations during consensus sampling for the more complex models can probably counteract this effect.

In contrast to the robust behavior of the parametric models, our implementation of the GP model quickly diverges with an increasing number of outliers as we did not model uniform outliers explicitly in this model. Note that there are several extensions to the GP framework that make the GP model robust against outliers, for example mixture noise models, Monte Carlo sampling methods, and iterative expectation maximization (Kuss, 2008).

For computing the data likelihood, all models estimate the outlier ratio γ of the training data. Figure 4.22 depicts the estimated outlier ratio versus the true outlier ratio. While the models using MLESAC estimators perfectly match the true outlier ratio, the GP model grossly overestimates the true value, as a result of its sensitivity to outliers in general: under the GP model, the training data appears to contain much more outliers than are actually present because it fails to recover the correct underlying structure.

Convergence of Model Estimation

We also evaluated the convergence behavior of our estimators with respect to the number of training samples on noisy but otherwise outlier-free data ($\sigma_{\mathbf{z},\text{pos}} = 0.05\,\text{m}$ and $\gamma = 0.0$). As expected, the sample error between the estimated and the true model converges to zero within the first 10 to 20 training samples, see Figure 4.23. Note that the rigid model needs at least

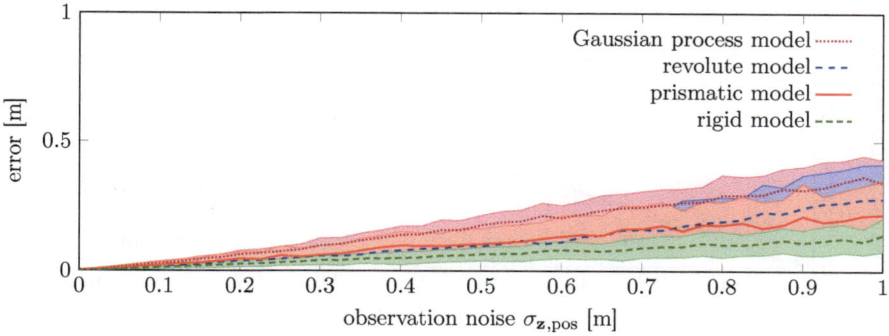

Fig. 4.20 Evaluation of the model prediction error on synthetic data for different noise values $\sigma_{\mathbf{z},\text{pos}}$. The lines correspond to the mean of 100 independent runs, the shaded areas mark a single standard deviation.

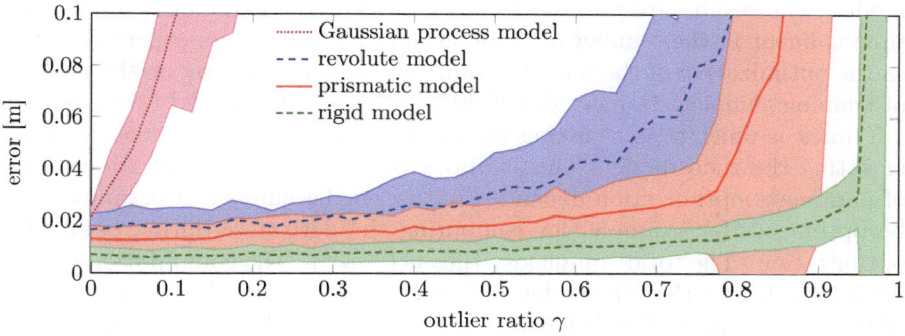

Fig. 4.21 Evaluation of the model prediction error on synthetic data for different outlier ratios γ.

one sample, for the prismatic model at least two, and for the revolute and the GP model at least three samples for training. The error of the rigid model falls below the observation noise already after the first sample, the prismatic and the GP model after three samples, the revolute model after six. The revolute model is thus the most sensitive to small training sets. We attribute this to the fact that it has the highest number of free parameters. When more training data is available, the GP model converges the slowest of all models: after 100 training samples, its error is still 2.5 times higher than the error of the rigid model. The revolute model has an error 2.1 times as high as the rigid model, and the prismatic model 1.5 times as high. We explain this effect by the high flexibility of the GP model. In general, more complex models have more free parameters to be estimated, are thus more sensitive to noise in the training data and thus take longer to converge.

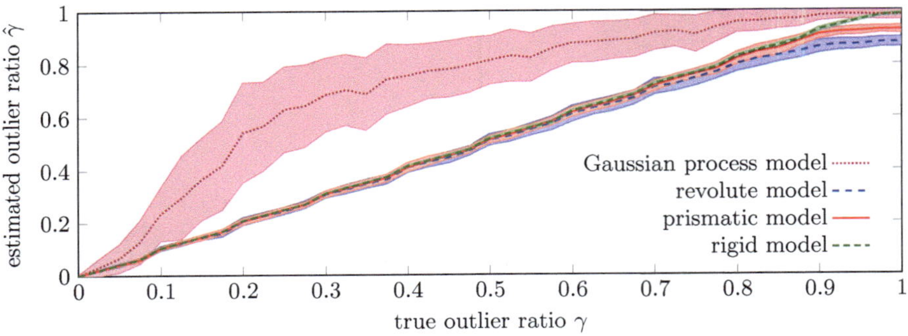

Fig. 4.22 Comparison of estimated outlier ratio versus true outlier ratio on synthetic data.

Further, we measured the computational time required for estimating the model. The results are given in Figure 4.24. For the parametric models the time is linear in the number of training samples. This is an expected result, as the optimization of the model parameters depends linearly on the number of training samples. In more detail, fitting the revolute model takes roughly 1.5 times as much time as fitting the prismatic model and 2.1 times as much as fitting the rigid model. This is also expected, as the computational load of parameter optimization depends linearly on the number of parameters to be optimized, for example, for computing the derivative during parameter optimization. The total complexity thus is $O(nk^2)$. The computational complexity for fitting the GP model, however, depends cubically on the number of training samples. This is because of the inversion of the covariance matrix which is an $O(n^3)$ operation.

In summary, we showed in this section that our estimators are robust against noise and, with some restrictions for the GP model, are robust against outliers. Further, we showed that they converge quickly with the number of training samples. From these results, we conclude that models learned from ten or more training samples are already significantly more accurate than the observation noise and can thus provide valuable information about an articulated object to a robot. The parametric models are robust to up to 50 % outliers in the training data and are thus well suited for robots with unreliable sensors. Finally, the computation time required for learning all models is below 0.25 s for training sequences consisting of up to 30 observations and thus our approach is suitable for online control.

Convergence of Model Selection

In a second series of experiments on synthetic data, we evaluated the model selection behavior with respect to the number of training samples and the

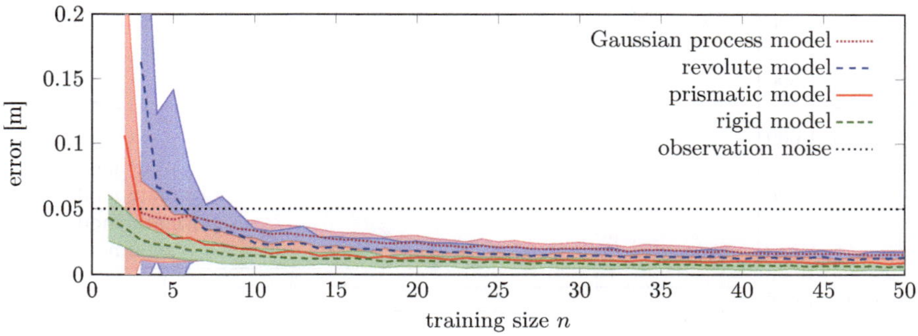

Fig. 4.23 Evaluation of the prediction error with respect to the number of training samples. The average positional prediction error decreases quickly with number of training samples n for all estimators.

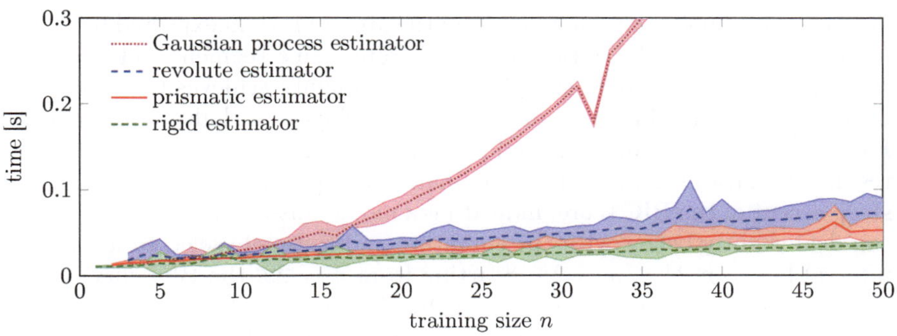

Fig. 4.24 The average runtime of all model estimators. The parametric model estimators run in $O(nk^2)$ time, while the GP model has a runtime of $O(n^3)$.

assumed observation noise level. For this analysis, we sampled noisy observations, fitted the candidate models, and selected the best model. We repeated all experiments for 10 independent runs and evaluated the mean and the standard deviation. In the following, we present our results exemplary for data sampled from the simulated roll-up door (see Figure 4.19d).

We present the training samples in the order of their configuration $\mathbf{q}$, i.e., we first presented data sampled from the vertical line segment, then additionally from the quarter circle, and finally also from the horizontal line segment as visualized in Figure 4.19d. Figure 4.25 shows the result of the first experiment, the relation between the BIC score and the number of training samples. Remember that the model with the lowest BIC score gets selected. Initially, after the first training sample is presented, only the rigid model can be fitted. Starting from the second sample, the prismatic model explains the

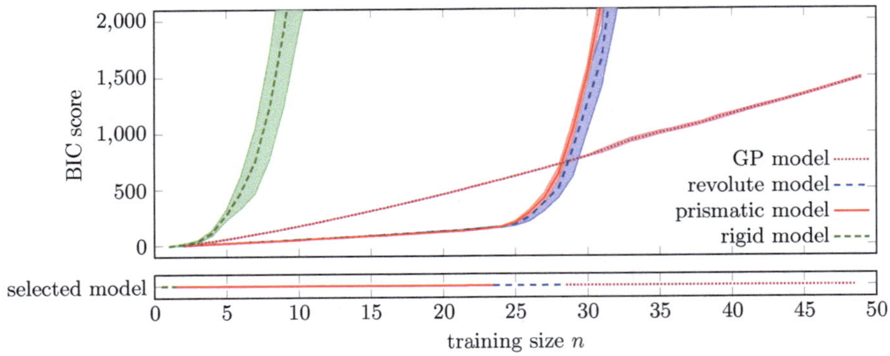

Fig. 4.25 BIC score and resulting model selection after observing a sequence sampled from the roll-up door.

data best until points are sampled from the arc. This is an expected behavior, as the first 40 % of the data truly come from a line segment. The score of the revolute model is only slightly worse; it roughly achieves the same data likelihood, but has three parameters more, leading to a slightly higher score. For $20 \leq n \leq 30$, training samples originate from the quarter circle. As a result, the data likelihood of the prismatic model drops quickly, leading to a steep ascent of the BIC score, indicating a model mis-fit. The revolute model can deal slightly better with the curvature and thus gets selected for a short interval between $24 \leq n \leq 28$. Starting from $n = 29$, the BIC curve of the revolute model surpasses the curve of the GP model and the GP model gets selected. Note that the number of parameters of the GP model grows linearly with the number of training samples as it needs to store all training data. However, its data likelihood stays almost constant, and, as a result, the BIC score grows linearly with the number of training samples.

Furthermore, we investigated the influence of the assumed observation noise $\Sigma_{\mathbf{z}}$ on the model selection process. When the number of training samples n is kept fixed, then higher noise favors the selection of simpler models, and vice versa. Figure 4.26 illustrates this dependency: for $n = 50$ and $\sigma_{\mathbf{z},\text{pos}} \leq 0.02$, the GP model is selected. Between $0.02 \leq \sigma_{\mathbf{z},\text{pos}} \leq 0.2$, the revolute model yields the best trade-off between model complexity and data likelihood. Above $0.2 \leq \sigma_{\mathbf{z},\text{pos}}$, the rigid model best explains the observations as the noise level of this magnitudes hides the underlying model.

From this set of experiments on synthetic data, we conclude that the proposed estimators are robust against normally distributed noise and that the MLESAC-based estimators are additionally robust against uniformly distributed outliers. We illustrated the effect of the number of training samples and noise assumptions on the model selection process and showed that model selection asymptotically converges to the correct model with a convergence rate that depends on the observation noise.

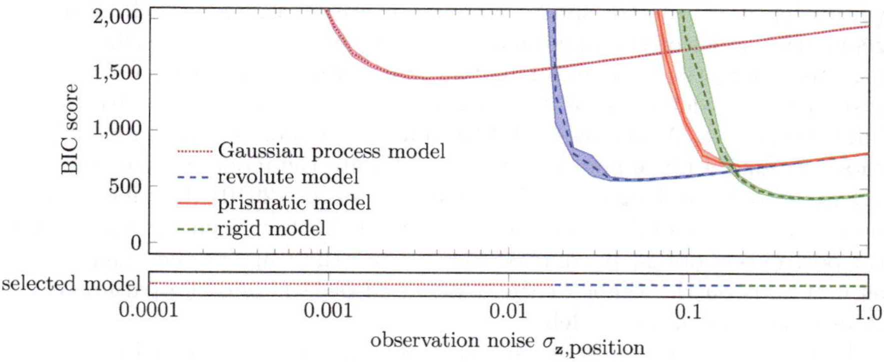

Fig. 4.26 BIC score as a function of the assumed observation noise. A low noise assumption favors the selection of more complex models, and vice versa.

4.5 Related Work and Discussion

Several researchers addressed the problem of operating articulated objects with robotic manipulators. A large number of these techniques focused in particular on the problem of handling doors and drawers in domestic environments (Klingbeil et al., 2009; Kragic et al., 2002; Meeussen et al., 2010; Petrovskaya and Ng, 2007; Parlitz et al., 2008; Niemeyer and Slotine, 1997; Andreopoulos and Tsotsos, 2008; Rusu et al., 2009; Chitta et al., 2010). Meeussen et al. (2010) described an integrated navigation system for mobile robots including vision- and laser-based detection of doors and door handles that enabled the robot to successfully open doors using a compliant arm. Diankov et al. (2008) formulated door and drawer operation as a kinematically constrained planning problem and proposed to use caging grasps to enlarge the configuration space. They demonstrated their approach on an integrated system that successfully performed a wide variety of different fetch-and-carry tasks (Srinivasa et al., 2010). Wieland et al. (2009) combined force and visual feedback to reduce the interaction forces when opening kitchen cabinets and drawers. The majority of these approaches, however, implicitly assumes that the kinematic model of the articulated object is known.

Katz and Brock (2008) enabled a robot to learn the kinematic model of a articulated object using vision by interacting with it on a table. Jain and Kemp (2009b) presented an approach that enabled a robot to estimate the radius and location of the axis for revolute joints that move in a plane parallel to the ground while opening doors and drawers using equilibrium point control. In contrast to our work, both approaches assume planar objects and learn only two-dimensional kinematic models.

There are several approaches where tracking articulated objects is the key motivation. Krainin et al. (2010), for example, developed an approach for tracking articulated objects such as a manipulator using a depth camera with a texture projector similar to ours. However, this approach requires a geometric model of the manipulator. Kragic et al. (2002) described an integrated navigation system for mobile robots which includes a vision-based system for the detection of door handles and enables the robot to successfully open doors. Anguelov et al. (2004) modeled doors as line segments that rotate around a hinge and used EM to estimate the model parameters from 2D range data and images. Nieuwenhuisen et al. (2010) described an approach where a mobile robot increases its localization accuracy by learning the positions of doors. In contrast to our work, these approaches make strong assumptions on the shape of the articulated objects and the parameterizations of the kinematic models.

Estimating kinematic structure from observations was studied in the field of computer vision, however, without subsequently using these models for robotic manipulation. Taycher et al. (2002) addressed the task of estimating the underlying topology of an observed articulated body. They focused on the recovery of the object topology rather than on learning accurate generative models. Also, compared to their work, our approach can handle links with more complex link models, including links with multiple DOFs and nonparametric models. Kirk et al. (2004) extracted human skeletal topologies using 3D markers from a motion capture system, however assuming that all joints are revolute. Yan and Pollefeys (2006) presented an approach to learn the structure of an articulated object from feature trajectories under affine projections. Other researchers addressed the problem of identifying different object parts from image data. Ross et al. (2010) used multi-body structure from motion to extract rigid parts from an image sequence and fitted motion models to these links using maximum likelihood learning. They also used graphical models to represent the kinematics, but introduced special variables to indicate whether a particular edge in the graphical model is active or inactive. This transforms the structure search into a parameter optimization problem. However, this might result in local minima and the resulting models are not accurate enough to be used by manipulation robots.

Kemp (2005) aimed at finding an assignment of body parts to internal sensors of a motion tracking suit by minimizing a traveling salesman problem with a suitable cost function. Although learning the structure of general Bayesian networks has been proven to be NP-complete (Chickering, 1996), many approximate methods have been proposed that can solve the structure search problem efficiently. Such methods include greedy structure search, iterated hill climbing, genetic algorithms and ant colony optimization (Chickering, 2002; Daly and Shen, 2009). In some cases, the size of the search space can be reduced significantly by evaluating a number of statistical independence tests (Margaritis and Thrun, 1999; Bromberg et al., 2009).

In contrast to all of the above work, we provide with our approach a complete probabilistic framework that enables a robot to learn accurate, three-dimensional kinematic models of articulated objects. Furthermore, our framework provides the kinematic structure and degrees of freedom, and allows a robot to use the learned models to operate articulated objects.

4.6 Summary

In this chapter, we presented a complete probabilistic framework for learning kinematic models of articulated objects. We learn both parametric and non-parametric models to describe the geometric relationship between connected object parts and consistently apply Bayesian model comparison to select the best models and infer the kinematic structure. In extensive experiments carried out on real robots and in simulation, we demonstrated that our approach is efficient and provides accurate kinematic models from noisy observations. Furthermore, we showed that our approach is applicable to a wide range of articulated objects and can be used in conjunction with a variety of different sensor modalities. Our approach enables mobile manipulators to learn accurate kinematic models of unknown articulated objects, operate them reliably, and can improve their learning performance by exploiting prior experience.

Chapter 5
Vision-Based Perception of Articulated Objects

The probabilistic framework developed in the previous chapter enables a manipulation robot to learn accurate kinematic models of articulated objects. As input, our framework requires a sequence of pose observations of the articulated object. We implemented the perception in the previous chapter using visual markers or by directly recording the end effector trajectory while the robot was manipulating the articulated object. For the daily use in domestic environments, however, both options are not satisfactory: clearly, it is neither desirable to augment all furniture with visual markers nor to guide a robot manually to the handles of all relevant objects.

In this chapter, we investigate how a robot can perceive the poses of cabinet doors and drawers without requiring artificial markers. We use a stereo camera system with an additional projector to generate dense depth images. As cabinet fronts appear in the depth images as rectangles, we develop a sampling-based approach that efficiently detects and tracks rectangles in these images. Our perception algorithm can be adapted to the computational capabilities of the robot as it allows to adjust the number of pose candidates per frame. After a sequence of pose observation has been acquired, the robot can learn the kinematic model of the articulated object using the approach presented in the previous chapter. In particular, our approach allows a robot (1) to infer the model class of the tracked object, (2) to estimate its current configuration, and (3) to make predictions about future configurations. In our experiments, we demonstrate that robots using our approach can learn accurate kinematic models of cabinets without requiring artificial markers in the environment. This is an important prerequisite for using mobile manipulation robots in domestic environments. Furthermore, we provide a detailed error analysis on our pose estimator based on ground truth data obtained in a motion capture studio.

The experimental setup is depicted in Figure 5.1 in which the robot observes the motion of a cabinet drawer. Our approach segments the depth image generated by the active stereo system and iteratively fits rectangles as illustrated in Figure 5.2a. By tracking the pose observations of the articulated object over time, the robot can learn the corresponding kinematic model and use it to predict future configurations of the object as visualized in Figure 5.2b.

J. Sturm: *Approaches to Probabilistic Model Learning*, STAR 89, pp. 113–124.
DOI: 10.1007/978-3-642-37160-8_5

Fig. 5.1 Experimental setup: the robot tracks the pose of a drawer in a sequence of dense depth images and infers its kinematic model.

This chapter is organized as follows. In Section 5.1, we develop our approach on detecting and tracking articulated objects in depth images. In Section 5.2, we analyze the properties of our approach in experiments carried out on a real robot in a domestic environment. Further, we used a motion capture studio to evaluate the detection rate and pose accuracy of our approach. Finally, we conclude this chapter with a discussion of related work in Section 5.3.

5.1 Marker-Less Pose Estimation

We assume that we obtain in each frame a dense depth image $D \in \mathbb{R}^{640\times480}$ from the stereo camera system, that contains for each pixel (u, v) its perceived disparity $D(u, v) \in \mathbb{R}$. The relationship between 2D pixels in the disparity image and 3D world points is defined by the projection matrices of the calibrated stereo camera and can be calculated by a single matrix multiplication from the pixel coordinates and disparity. For acquiring dense depth images, we use the stereo camera system in conjunction with a texture projector similar to the one used on the PR2 robot (Konolige, 2010).

5.1.1 Fast Processing of Depth Images

We apply the RANSAC algorithm to segment each depth image into planes, i.e., we iteratively sample three pixels from the depth image, estimate the corresponding plane coefficients $\mathbf{z}_{\text{plane}} \in \mathbb{R}^4$, and count the inliers of that plane. We define the plane to comprise all pixels that are within a certain distance L of the plane, i.e.,

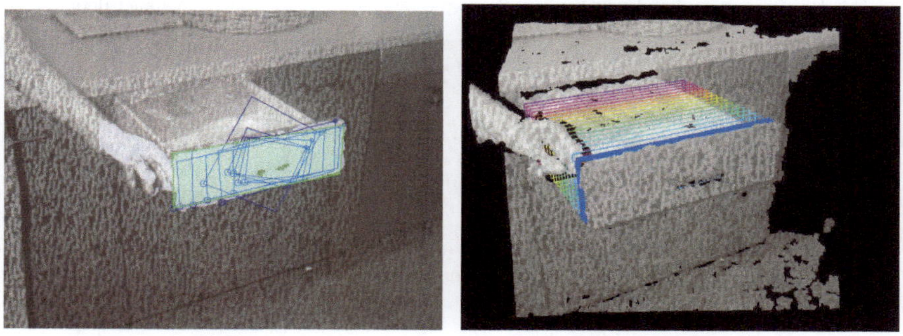

(a) (b)

Fig. 5.2 Illustration of the processing steps of the proposed approach. (a) The pose of the rectangle is iteratively optimized. (b) By tracking the poses over multiple frames, the kinematic model can be learned.

$$\|\mathbf{z}_{\text{plane}}(x\ y\ z\ 1)^T\| \leq L. \tag{5.1}$$

In general, L depends on the particular noise level of the camera – in our case, we used $L = 0.02\,\text{m}$. We repeat this process of plane candidate generation until we find a plane with a high enough support, or we exceed a given number of iterations. We select the plane with the most inliers and subtract the corresponding inliers from the point cloud. Subsequently, we apply the same strategy to the remaining points in the cloud, until no more points remain or the maximum number of planes has been reached.

In contrast to typical approaches to RANSAC-based plane fitting which always assign pixels to one plane, our masks allow points to belong to several planes at the same time. This is useful, as the infinite planes determined via RANSAC always intersect with the subsequent (less significant) planes, thereby cutting out points that make detection of contiguous rectangles more difficult in the next step of the perception process.

For a visualization of the result, see Figure 5.3. In this example, our algorithm automatically segmented three planes from a depth image of a cabinet door. For each plane, we create an image mask M with class labels for the pixels in the depth image, i.e., $M \in \{\text{in-plane}, \text{free}, \text{occluded}, \text{unknown}\}^{640\times 480}$, with

$$M(u,v) = \begin{cases} \text{in-plane} & \text{if} \quad \|\mathbf{z}_{\text{plane}}(x\ y\ z\ 1)^T\| \leq L \\ \text{free} & \text{if} \quad \mathbf{z}_{\text{plane}}(x\ y\ z\ 1)^T > L \\ \text{occluded} & \text{if} \quad \mathbf{z}_{\text{plane}}(x\ y\ z\ 1)^T < -L \\ \text{unknown} & \text{if missing value} \end{cases}. \tag{5.2}$$

Here, "in-plane" indicates that the pixel belongs to the plane for which the mask M is computed. In contrast to that, "free" indicates that the observed pixel lies behind the plane and "occluded" that a pixel in front of the plane

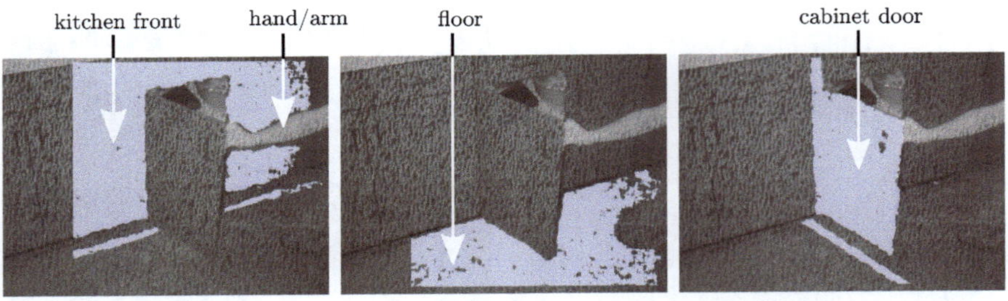

(a) first plane (b) second plane (c) third plane

Fig. 5.3 This figure shows the segmentation of a depth image into the three most prominent planes with RANSAC.

has been observed which occludes the plane. "Unknown" means that no depth information is available for that pixel.

The next step is to find rectangles in the segmented planes. A rectangle in 3D space has 8 degrees of freedom: its position, its orientation and its dimensions (3+3+2). After the plane segmentation, we have already fixed 3 DOFs, so that we need to find the remaining 5 DOFs. We apply an iterative fitting approach here. We start with a sampled candidate rectangle and optimize its pose and size iteratively using an objective function g.

For creating an initial rectangle candidate, we sample a random point from the plane and sample the other DOFs from a prior distribution. The objective function g is based on the average cost of the pixels inside the rectangle $\mathbf{z}_{\text{rect}} \in \mathbb{R}^8$,

$$g(\mathbf{z}_{\text{rect}}) := -\frac{1}{|\text{pixels}(\mathbf{z}_{\text{rect}})|^{1+\alpha}} \sum_{\text{pixels}(\mathbf{z}_{\text{rect}})} \text{cost}(M(u,v)). \tag{5.3}$$

The parameter α (that we empirically chose around $\alpha = 0.05$) makes g favor larger rectangles over smaller ones.

Finding a good cost metric cost, in particular for occluded and unknown pixels, is non-trivial. If chosen too low, the greedy search converges on too large rectangles, while a too high cost increases the amount of local maxima in g and in turn leads to the detection of partial rectangles in the presence of occlusions or unknown pixels (see Figure 5.4). Empirically, we found a cost value of cost(occluded) = cost(unknown) = 0.2 to be working well for our data, but the choice of this parameter depends in principle on the ratio of occluded and unknown pixels in the scene, and, therefore, needs to be adapted for different environments.

5.1.2 Pose Estimation

In each iteration, we now individually optimize every DOF of the rectangle. We apply a small set of discrete changes to each DOF and evaluate the ob-

(a) cost too high (b) cost too low (c) good segmentation

Fig. 5.4 Illustration of the effect of the cost parameter for unknown and occluded pixels. (a) Cost too high (1.0). (b) Cost too low (0.0). (c) Good (0.2).

jective function on $\mathbf{z}'_{\text{rect}}$. If $g(\mathbf{z}'_{\text{rect}}) > g(\mathbf{z}_{\text{rect}})$, we continue with the improved parameter set. When this greedy search converges (or we reach the maximum number of iterations), we need to evaluate the quality of the found match. In preliminary experiments, we found that the value of the objective function was not sufficient for discrimination of false and true positives.

Therefore, we decided to evaluate the rectangle candidate $\mathbf{z}_{\text{rect}}$ using two measures that are inspired from statistical classification theory and that have a natural interpretation. First, we evaluate the precision $r_{\text{precision}}$ of the rectangle candidate as the ratio of detected pixels and all pixels in the found rectangle. Second, we evaluate the recall r_{recall} as the ratio of pixels in the found rectangle versus the pixels in the selected plane. For both measures, we use our cost functions to weight occluded and unknown pixels accordingly:

$$r_{\text{precision}}(\mathbf{z}_{\text{rect}}) := \frac{\sum\limits_{\text{pixels}(\mathbf{z}_{\text{rect}})} 1 - \text{cost}(M(u,v))}{|\text{pixels}(\mathbf{z}_{\text{rect}})|} \tag{5.4}$$

$$r_{\text{recall}}(\mathbf{z}_{\text{rect}}) := \frac{\sum\limits_{\text{pixels}(\mathbf{z}_{\text{rect}})} 1 - \text{cost}(M(u,v))}{\sum\limits_{\text{pixels}(\mathbf{z}_{\text{plane}})} 1 - \text{cost}(M(u,v))}. \tag{5.5}$$

Empirically, we found that a good condition for thresholding is to require that both ratios are above 0.7, which removes most of the false positives.

An example of the iterative pose fitting is given in Figure 5.5a: the rectangle candidate started in the lower left of the door and iteratively converged to the correct pose and size of the door. The candidate is accepted, because both ratios $r_{\text{precision}}$ and r_{recall} have high values. The greedy search however can get stuck in local maxima. In the example depicted in Figure 5.5b, the hand is also part of the drawer front plane and the candidate rectangle converged to a rectangle that fits to some extend the hand. Our algorithm

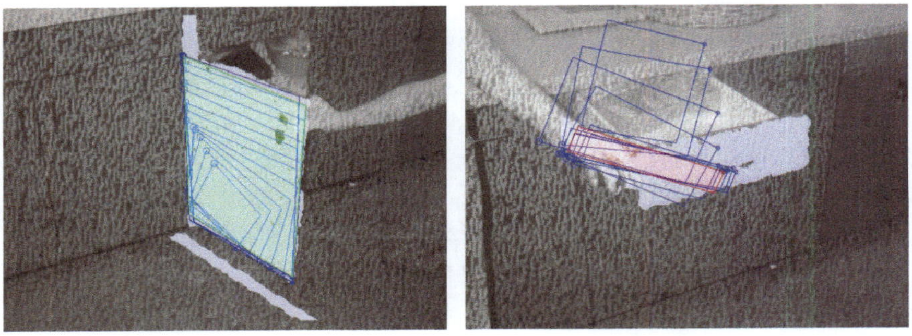

(a) (b)

Fig. 5.5 Illustration of the iterative matching and filtering of rectangles in depth images. (a) Accepted sample. (b) Rejected sample.

rejects this candidate rectangle because it does not contain the majority of pixels in the plane, i.e., r_{recall} takes a low value.

We deal with the problem of local maxima by starting from several rectangle candidates. In this sense, our algorithm is probabilistically complete, as we would find any visible rectangle in the limit with probability 1 given enough candidates. In practice, we chose a fixed number of m samples per plane.

5.1.3 Pose Tracking

In the remainder of this chapter, we drop the subscript in $\mathbf{z}_{\text{rect}} = \mathbf{z}$ to improve readability. The rectangle detector described in the previous section gives us per frame between zero and $r \cdot s$ observations of rectangles (r rectangle candidates in s planes), which need to be integrated into consistent tracks. Checking whether two rectangles $\mathbf{z}_i$ and $\mathbf{z}_j$ are similar requires to take the ambiguity in the representation into account: the same rectangle can be described by eight different parameter vectors (depending on the choice of the corner of origin and the choice of the front or back side of the rectangle). As a result, we obtain an integrated sequence

$$\mathcal{D}_{\mathbf{z}}^t = \left\langle \mathbf{z}^1, \ldots, \mathbf{z}^n \right\rangle, \tag{5.6}$$

when after t time steps n rectangles have been observed.

In our implementation, we check whether a new observation $\mathbf{z}^{\text{new}}$ (under consideration of the above-mentioned ambiguities) is close to an existing track $\mathcal{D}_{\mathbf{z}}^i$. Then it is either appended to that track $\mathcal{D}_{\mathbf{z}}^{i,t+1} := \left\langle \mathbf{z}^1, \ldots, \mathbf{z}^n, \mathbf{z}^{n+1} \right\rangle$, or a new track is initialized $\mathcal{D}_{\mathbf{z}}^{\text{new},t+1} := \left\langle \mathbf{z}^{n+1} \right\rangle$. For deciding whether a disambiguated observation is close enough to an existing track, we use fixed thresholds on pose change and considered also the uncertainty in the estimate of the object size.

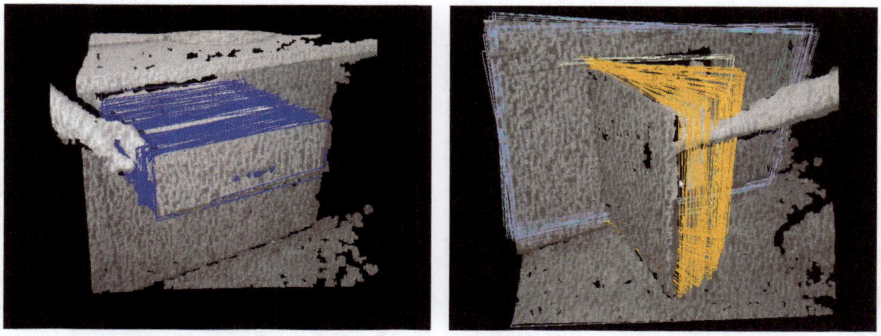

(a) (b)

Fig. 5.6 Examples of the observed tracks of a (a) cabinet drawer and (b) cabinet door, respectively.

Figure 5.6 shows two examples of the resulting tracks obtained when observing a drawer and a door. In Figure 5.6a, only a single track for the drawer was created, while in Figure 5.6b two tracks for the door (yellow and orange) were created and not yet merged, as well as two stationary tracks (purple and cyan) were instantiated corresponding to the background. In this experiment, our model selection framework selected a rigid link for these additional trajectories.

With this mentioned approach, we obtain for each moving part i in the scene a trajectory of observations $\mathcal{D}^i_{\mathbf{z}}$. This trajectory can directly be used to learn the kinematic model as described in Chapter 4.

5.2 Experiments

We conducted two sets of experiments. The goal of the first set of experiments was to evaluate the performance of our rectangle detector, pose estimator, and tracker. In the second set, we demonstrate that our approach enables a robot to successfully learn kinematic models of articulated objects.

5.2.1 Evaluation of Detection Rate and Pose Accuracy

To evaluate the performance of our approach to marker-less perception, we placed a PR2 robot inside a motion capture studio and recorded time-synchronized stereo images with the robot and ground truth pose information from the motion capture system. For our experiments, we used an unmounted drawer of a typical office cabinet that we equipped with 5 tracking LEDs. We recorded several sets of logfiles containing 19,412 stereo images in a large variety of different poses. Figure 5.7 shows a typical point cloud (gray) from

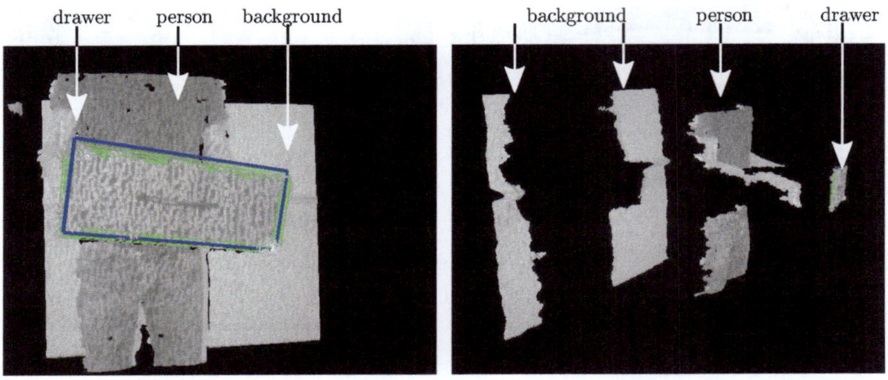

(a) frontal view (b) side view

Fig. 5.7 Illustration of the ground truth evaluation of our system. The blue rectangle corresponds to the ground truth location reported by the motion capture system, the green rectangles are our estimates.

this dataset including an overlay of the detected pose (green) and the ground truth pose (blue). For evaluation, we used the ground truth pose information for checking whether the drawer was fully visible in both camera images. If so, we checked whether a rectangle was detected and measured its positional and orientational error.

As a first result, we found that the drawer was correctly detected in more than 75 % of the images up to a distance of 2.2 m from the camera (see Figure 5.8a, red curve). We attribute the drop of the detection rate after 2.2 m to the maximum distance of the texture projector: when the object is too far away from the camera, too many dropouts (missing values in the depth image) occur and the depth image is no longer dense. The range of the stereo camera system can either be increased by using a stronger projector lamp, or by using other reconstruction techniques (Brox et al., 2010). With the distance of the drawer to the camera, also the number of planes increases that need to be searched before the drawer is detected. This is an expected result since the drawer appears smaller in the depth image the further it is away (same figure, blue curve).

Further, we evaluated the positional and orientational error by comparing the detected pose using our approach with the ground truth obtained from the motion capture studio. Figure 5.8b gives the result. We found that the average position error of the estimator was on average below 0.015 m. It also was almost independent of the actual distance to the camera. The same holds for the orientation error, that was on average below 3°.

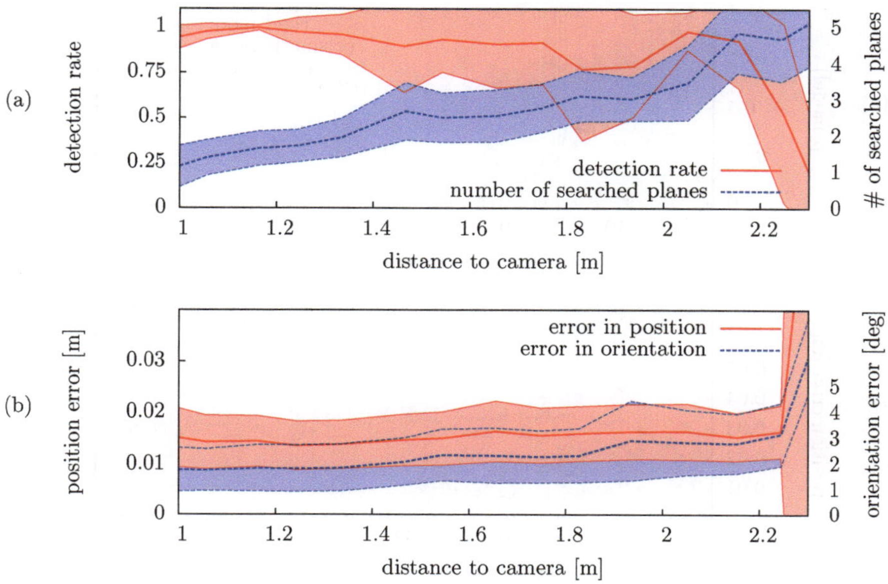

Fig. 5.8 Evaluation of the detector with respect to the distance to the camera. The line corresponds to the mean, and the shaded area to a single standard deviation. (a) Detection rate and number of searched planes. (b) Position and orientation error.

5.2.2 Kinematic Model Learning

To evaluate the quality of the learned kinematic models, we recorded detailed logfiles of both a door (0.395 m × 0.58 m) and a drawer (0.395 m × 0.125 m) of a typical kitchen interior that we repeatedly opened and closed. We recorded a total of 1,023 and 5,202 images. From these logs, we sampled uniformly around 60 images in correct temporal order, and ran our detector and tracker as described in Section 5.1 on the down-sampled logfile. We trained the candidate models on the resulting tracks and evaluated the outcome of the model selection. We repeated this evaluation 50 times to verify the robustness of our approach. Figure 5.9 shows the result of the drawer dataset and Figure 5.10 for the door. For the datasets of both objects, we found that for the first 10 observations, mostly the rigid model was selected, as no substantial motion of the drawer or door was yet detected. With an increasing number of pose observations, the predictive error of the rigid model grows, while the predictive errors of the prismatic and the revolute models still remain low. After 30 observations, model selection has converged in all cases to the true model, i.e., the prismatic model for the drawer and the revolute model for the door. For the drawer model we measured prediction errors of 0.01 m and 7°; and 0.01 m and 3.5° for the door. In additional experiments, we found that

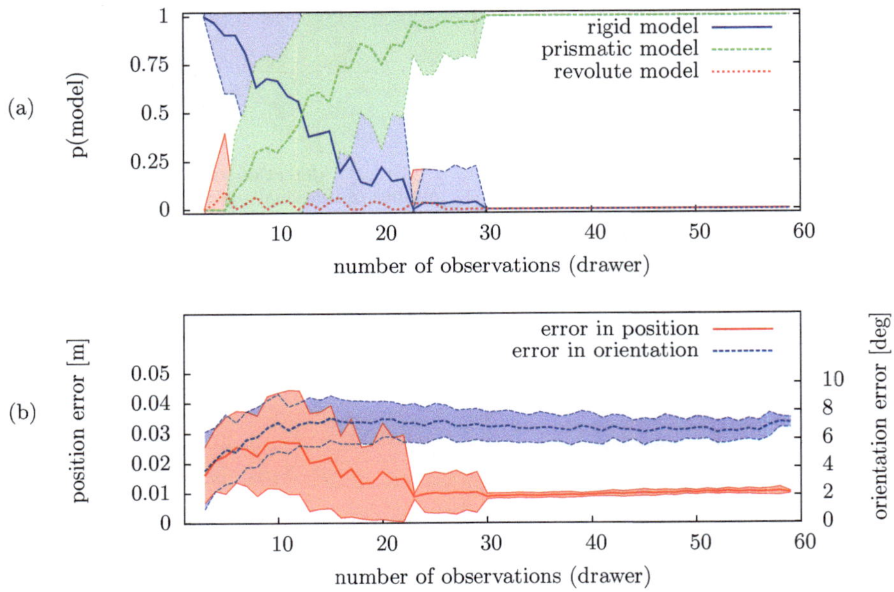

Fig. 5.9 Evaluation of model learning and selection for a cabinet drawer with respect to the number of training samples. (a) Posterior probability of candidate models. (b) Prediction error of the learned model. The line corresponds to the mean, the shaded area to the standard deviation.

the pose estimates of the drawer are more sensitive to distortions around the horizontal axis because of its small height (0.12.5 m) of the drawer.

In our current, un-optimized implementation, the plane extraction takes on average 0.845 s on a single 2 GHz Pentium core. Creating the image mask of each plane takes approximately 0.008 s. Sampling a rectangle candidate from the mask takes 0.010 s, optimizing the pose around 0.313 s, and finally checking the precision and recall of the candidate consumes another 0.0023 s.

In additional experiments, we validated our approach on large number of different doors and drawers in two different kitchens. Also, we successfully tested the detector on a small office pedestal with three drawers of different size, a fuse door, and a fire extinguisher door in the wall.

From these experiments, we conclude that our approach is reliable, i.e., it detects doors and drawers up to a distance of 2.2 m in more than 75 % of the frames. Furthermore, we demonstrated that the pose accuracy of our system is comparable to a marker-based system, i.e., the average error in our experiments was below 0.015 m and 3°. Finally, we showed that at this accuracy, the resulting pose trajectories can be used to learn accurate kinematic models with prediction errors below 0.01 m and 7°.

(a)

(b)

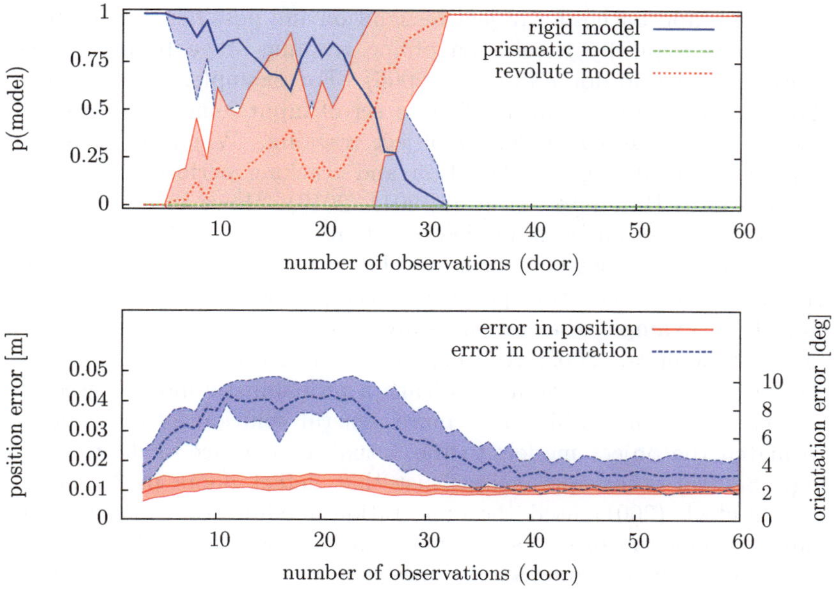

Fig. 5.10 Same as Figure 5.9, but for the door dataset. The plots show the evaluation of the learned kinematic model. (a) Posterior probability. (b) Prediction error.

5.3 Related Work and Discussion

For our application, we require accurate and dense point clouds of the scene at video frame rates. Flash ladars (Anderson et al., 2005) often have poor depth and spatial resolution and have non-Gaussian error characteristics that are difficult to deal with. Line stripe systems (Curless and Levoy, 1995; Quigley et al., 2009) have the requisite resolution but neither achieve 15 Hz operation nor deal with moving objects. Stereo systems that employ matching algorithms to produce dense results (Brox et al., 2010; Konolige, 1997; Wedel et al., 2008) can be a suitable sensor for our application. However, passive stereo suffers from the problem of *dropouts*, i.e., areas of low texture that cannot be matched correctly. An interesting technology is to use structured light (Nishihara, 1984; Lim, 2009) either with monocular or stereo cameras. Recently, structured light systems have become available at comparatively low prices providing both depth and color images such as the Kinect sensor from Microsoft (Machline et al., 2010; Fox and Ren, 2010). For our work, we used a compact projector for active stereo with a fixed, random pattern developed by Konolige (2010). It provides a texture for stereo that produces excellent error characteristics at distances up to 3 meters, even for surfaces with low reflectivity.

General approaches for object recognition and pose registration from camera images typically require an offline training stage to learn a suitable object model (Gordon and Lowe, 2006). For example, Collet Romea et al. (2011) extracts SIFT features from a set of input images and uses them to generate a sparse 3D model for object tracking. When additionally depth images are available, the detection and tracking performance can further be increased (Hinterstoisser et al., 2011, 2012). However, none of these approaches have so far been applied to articulation model learning.

Several approaches have been developed to detect and estimate the pose of room doors. For this purpose, strong appearance models can be assumed, for example, that all doors are vertical or the dimensions are known (Monasterio et al., 2002; Andreopoulos and Tsotsos, 2008). Murillo et al. (2008) proposed an approach to learn a probabilistic model based on color and shape features which is a more general solution as the robot is able to match the object models to the actual appearance of doors in a building. Several other approaches exist that detect doors in 2D laser scans. Yufeng et al. (2001) used the expectation-maximization algorithm to estimate the position of doors. Nieuwenhuisen et al. (2010) detected dynamic obstacles during mapping and fit lines to recover the positions of doors in the scene. Rusu et al. (2009) presented an integrated approach for door and door handle detection using a tilting laser scanner. Other approaches focus on the semantic interpretation of (static) 3D point clouds. Nüchter and Hertzberg (2008), for example, seek for a segmentation and labeling of 3D point clouds into object classes such as walls, doors, floors and ceilings.

In contrast to all of these approaches, we use dense depth images acquired from a stereo camera system to estimate and track the poses of articulated objects. Further, we use the resulting tracks in our framework to estimate accurate kinematic models of these objects.

5.4 Summary

In this chapter, we presented an approach that enables a robot to detect and track cabinet fronts from dense depth images. Our system segments depth images into planes and iteratively fits rectangles to them. By tracking the detections over multiple frames, the robot obtains pose sequences from which it can learn the kinematic model of an articulated object using our approach from Chapter 4. In extensive experiments, we evaluated the performance of our system in a motion capturing studio and found high detection rates and accurate pose estimation. Furthermore, we demonstrated in our experiments that a real robot using our approach could learn highly accurate kinematic models of various doors and drawers in domestic environments. We consider the ability to cope without artificial markers an important feature that significantly increases the usability of mobile manipulation robots in everyday life.

Chapter 6
Object Recognition Using Tactile Sensors

So far, we have considered monocular and stereo vision to perceive the state of the world. However, vision alone is in some cases not sufficient: for example, a robot that picks up an object from a box cannot see which object it grasps because the robot partially occludes its view with its own gripper. In particular for robotic manipulation tasks, tactile sensing provides another sensor modality that can reveal relevant aspects about the object being manipulated, for example, to infer its identity, pose, and internal state.

In this chapter, we develop an approach that enables a manipulation robot to identify an object using tactile sensing. In our concrete scenario, we consider a manipulation robot which has touch-sensitive sensor arrays installed in its finger tips, and we assume that these sensors provide low-resolution pressure images of the grasped objects. An example of such a *tactile image* is depicted in Figure 6.1. In this experiment, the robot grasped the handle of a coffee mug with a two-fingered parallel jaw gripper. As a result, the robot observes a diagonal stripe in the tactile images corresponding to the handle of the mug.

Our approach uses the bag-of-features model that we apply to object classification based on tactile images. First, the robot generates a suitable tactile feature vocabulary using unsupervised clustering from real data. Second, it learns a set of feature models (a so-called *codebook*) that encodes the appearance of objects in form of feature histograms. After training, a robot can use this codebook to identify the grasped object. Since the objects that we consider are typically larger than the sensor and consist of similar parts, the robot may need multiple grasps at different positions to uniquely identify an object. To reduce the number of required grasps, we apply a decision-theoretic framework that chooses the grasping location in such a way that the expected entropy of the belief distribution is minimized. In experiments carried out on a large set of industrial and household objects, we demonstrate that our approach enables a manipulation robot to discriminate various objects only by touch.

The remainder of this chapter is organized as follows: in Section 6.1, we present our approach on tactile object recognition based on the bag-of-features model. In Section 6.2, we extend this approach with a decision-theoretic framework to minimize the number of grasps required to uniquely

J. Sturm: *Approaches to Probabilistic Model Learning*, STAR 89, pp. 125–139.
DOI: 10.1007/978-3-642-37160-8_6

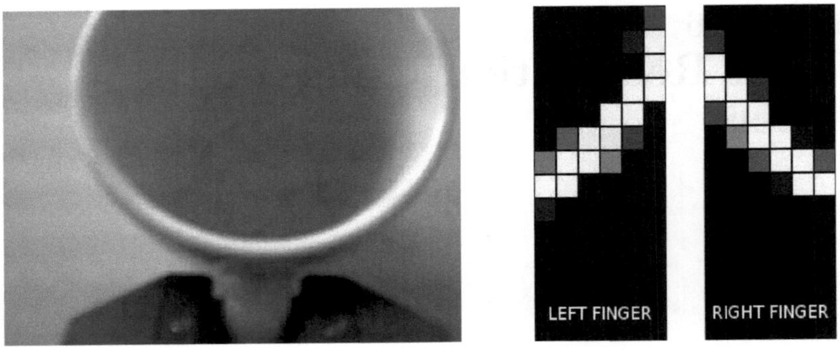

(a) (b)

Fig. 6.1 A manipulation robot with touch-sensitive finger tips grasps the handle of a coffee mug. (a) Visual image. (b) Tactile images showing the handle of the mug.

identify an object. We present the evaluation of our approach on a large set of different objects in Section 6.3. Finally, Section 6.4 concludes this chapter with a discussion of related work.

6.1 The Bag-of-Features Model

We developed our approach on a mobile manipulation robot that has a 1-DOF Schunk parallel gripper with two fingers as its end effector. Both fingers are equipped with a tactile sensor from Weiss robotics for gathering tactile images (Weiss and Wörn, 2005). Each tactile sensor consists of an array of 84 pressure-sensitive cells arranged in 6 columns and 14 rows with a size of 24 mm by 51 mm. The sensor principle is to measure the conductivity of an elastic rubber foam above a circuit board. When a force is applied to the rubber foam, the binding polymer gets compressed which lowers the electrical resistance of the material. The exact calibration of the sensor array turned out to be difficult in consequence of this sensor principle. However, we found that taking a reference measurement before the experiments (with no pressure on all cells) was a simple means to suppress memory effects of the rubber foam. Furthermore, we normalized all measurements to the sensor's maximum response, such that we obtained for each sensor array a measurement matrix $Z \in [0,1]^{6\times 14}$.

For generating training data, we assume that the robot interacts with various objects from a set of n different classes $C = \{1, \ldots, n\}$. During training, we assume further that the identity of an object is known to the robot, i.e., the corresponding class label $c \in C$ is given. The robot grasps each object repeatedly at different positions h along the object by slowly closing its gripper. In this work, we do not deal with the localization of the object but assume that a suitable coordinate system on the object is available to the

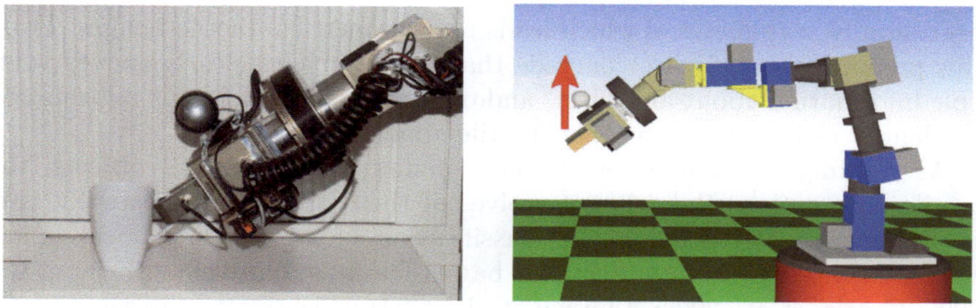

(a) (b)

Fig. 6.2 Experimental setup. The robot grasps an object at different positions. (a) Real robot grasping a mug. (b) Visualization of the planned trajectory.

robot, for example, from a vision system. As soon as the gripper has established a particular force on the object, the robot stops and records the tactile image of both of its fingers $Z^{\text{left}}, Z^{\text{right}} \in [0,1]^{6\times 14}$. Furthermore, the robot reads out the distance between its fingers which serves as an estimate of the object width w of the object at position h. As a result, we obtain from each grasp a tactile observation $\mathbf{z} = (Z^{\text{left}}, Z^{\text{right}}, h, w)$ that encodes the observed tactile images, grasping position, and corresponding object width. After N grasps, this yields a set $\mathcal{D} = \{(\mathbf{z}_i, c_i)\}_{i=1}^{N}$ of N of training tuples consisting of tactile observations annotated with the corresponding class labels. Our procedure for data acquisition is also illustrated in Figure 6.2. The robot approaches the object at a given height h and closes its gripper. When the gripper motion comes to a rest, it reads out the tactile sensor array and saves it to a database. After this, it opens again its fingers and approaches the next probing location.

We carried out this data acquisition procedure on a large set of objects. Figure 6.3 shows pictures of some of these objects and their corresponding tactile images. For these images, the gripper approached the object horizontally and grasped the objects at different heights. For example, consider the cuboid object in Figure 6.3a: As indicated by the black arrows, the parallel-jaw gripper embraces the object horizontally in the middle which yields a rectangular response on the tactile sensor array. This leads to the (mostly) white rectangle on the corresponding tactile images of both fingers. A more interesting tactile image is produced by the handle object in Figure 6.3d: The fingers touched the object approximately at the horizontal hole, which can be clearly recognized as a white circle in the tactile images. Other objects, like the toy figures in Figure 6.3i and Figure 6.3j lead to more complex patterns that are more difficult to interpret. Note that some of our objects are symmetrical which leads to mirror-symmetric tactile images at the left and right finger (e.g., the handle object in Figure 6.3d). However, other objects like the Goofy toy figure (Figure 6.3i or the mobile phone (Figure 6.3k) are horizontally asymmetrical which leads to asymmetrical tactile images. From this preliminary study, we conclude that tactile images clearly convey valuable information about an object, and our goal is to exploit this information to identify objects based on their tactile appearance.

As the finger of the robot is much smaller than all of our objects, the tactile observations the robot perceives of these objects are generally only partial views. To perform the classification based on these local image patches, we apply a variant of the bag-of-features approach (Zhang et al., 2007; Csurka et al., 2004; Agarwal et al., 2004) which have been successfully applied in the area of computer vision. The bag-of-features approach is appealing because of both its simplicity and power. The key idea is to describe the observations with a common vocabulary of features. For tactile perception, the vocabulary might include features such as "straight", "round", and "thin" observations. Given that the feature vocabulary is rich enough, the resulting feature histograms are well suited for object classification. For this purpose, a codebook needs to be learned that contains the feature histograms of the trained objects. The set of all feature histograms is called the *codebook*.

Figure 6.4 graphically illustrates the process of the codebook generation. In this example, we consider tactile observations of three different objects, i.e., bottle, beer glass, and coffee mug and five features. The features in of the vocabulary are "thick vertical", "medium vertical", "thin vertical", "horizontal", and "diagonal" indicated with the small patches below the columns of the histograms. By grasping each object at different positions (indicated by the highlighted rectangles on the objects), the robot estimates the occurrence frequency of each feature per object and learns the characteristic histograms. After training, the robot can use these histograms to uniquely identify the grasped object.

6.1.1 Unsupervised Creation of a Tactile Vocabulary

In practice, the appropriate vocabulary strongly depends on the objects that the robot is supposed to grasp so that pre-defined vocabularies will in general not suffice. Therefore, our goal is to create a suitable vocabulary automatically from the training data. Such a vocabulary can be created using k-means clustering as introduced in Chapter 2. This requires a distance metric of the tactile observations. A naive approach to compare two tactile images $R, S \in [0,1]^{6\times 14}$ is to compute the pixel-wise distance, i.e.,

$$d_1(R,S) := \sum_{x,y} |r_{xy} - s_{xy}|, \tag{6.1}$$

where r_{xy} and s_{xy} refer to the components of the tactile image matrices R and S, respectively. To allow for small translations of the object in the robot's fingers, we discount vertical shifts. We achieve this by defining a translation-invariant distance measure

Fig. 6.3 Visual and tactile images of some of the objects used in our experiments. The visual image is depicted at the top and the tactile images corresponding to the left and right finger at the bottom. The arrows in the visual image indicate the approach vector of the robotic gripper.

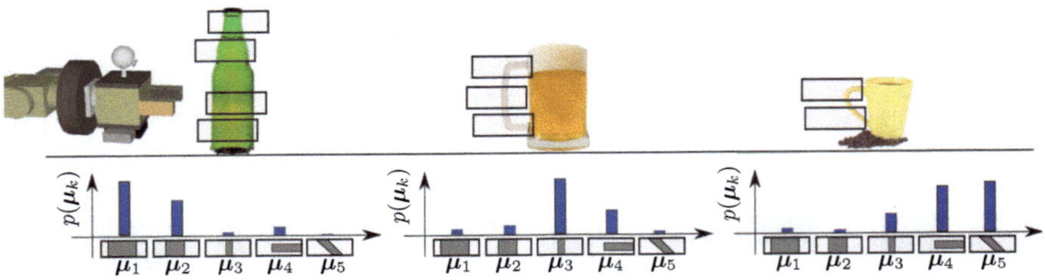

Fig. 6.4 Illustration of the bag-of-features approach with three objects described using five features. The learned histograms contain the occurrence frequencies $p(\boldsymbol{\mu}_k)$ of each tactile feature $\boldsymbol{\mu}_1, \ldots, \boldsymbol{\mu}_5$ on the object.

$$d_2(R, S) := \min_{\tau=-m,\ldots,m} (d_1(R, \mathit{shift}(S, \tau)), \tag{6.2}$$

where *shift* refers to a matrix operation that vertically shifts the content of the matrix S up- or downwards and m is the parameter that specifies the number of rows that the matrix can be shifted. Finally, we create a combined distance for two observations tuples $\mathbf{z}_1, \mathbf{z}_2$ comprising both fingers and the fingertip distance, i.e.,

$$\mathit{dist}(\mathbf{z}_1, \mathbf{z}_2) := \alpha \left(d_2(Z_1^{\text{left}}, Z_2^{\text{left}}) + d_2(Z_1^{\text{right}}, Z_2^{\text{right}}) \right) + (1-\alpha)|w_1 - w_2|, \tag{6.3}$$

where $\alpha \in [0, 1]$ is a weighting factor determining the influence of differences in tactile and finger tip distance. In order to circumvent scaling issues between tactile distances and finger distances, we normalize both of them to have unit variance on the training set. The result of this clustering step are k cluster centers denoted by $\boldsymbol{\mu}_1, \ldots, \boldsymbol{\mu}_k$ that form the tactile features of our vocabulary.

6.1.2 Learning the Feature Histograms

The next step is to learn the codebook, i.e., the set of feature histograms describing the objects. Each element of the codebook, or feature histogram, $\mathbf{h}^c$ represents to the probability distribution over features for a particular object c. Therefore, each histogram $\mathbf{h}^c$ has k bins, one for each tactile feature, i.e., $\mathbf{h}^c \in \mathbb{R}^k$. We denote the overall set of such histograms for the codebook by $H \in \mathbb{R}^{k \times n}$.

To learn this codebook, we initialize $\mathbf{h}^c = \mathbf{0}$ and update each bin h_i^c of $\mathbf{h}^c$ according to the observations $\mathbf{z}$ of object c in $\mathcal{D}_{\text{training}}$ by

$$h_i^c \leftarrow h_i^c + \exp(-\mathit{dist}(\boldsymbol{\mu}_i, \mathbf{z})/l), \tag{6.4}$$

for all $i \in \{1, \ldots, k\}$, where l is the length scale parameter in the observation distance space. After processing all observations, the individual $\mathbf{h}^c$ must be normalized so that the histogram sums to one.

The key idea of the codebook is to have a compact representation of the objects that allows us to efficiently compute the likelihood that a new observation is generated by touching a specific object c.

6.1.3 Object Classification

To compute the distribution over potential object classes based on an observation, we proceed as follows. By applying Bayes rule, we can write

$$p(c \mid \mathbf{z}) = \eta p(\mathbf{z} \mid c)p(c), \tag{6.5}$$

where η is a normalizing constant ensuring that the left-hand side sums up to one over all c. The term $p(c)$ is the prior over the object classes. In practice, this can be estimated from the training data or alternatively assumed to be uniformly distributed.

To compute the observation model $p(\mathbf{z} \mid c)$, we generate a histogram $\mathbf{h}^{\mathbf{z}}$ of a single observation $\mathbf{z}$ according to Eq. (6.4). As a result, we have two distributions over feature occurrences. With this, we can compute $p(\mathbf{z} \mid c)$ using a similarity measure between feature histograms of current observation and the histogram stored in the codebook.

There are multiple ways for computing the similarity between histograms. Among the popular measures for comparing histograms (Hetzel et al., 2001) are the histogram intersection, the χ^2 distance and the Kullback Leibler divergence (KLD). In preliminary experiments, we found that for our application, histogram intersection yielded the best recognition results. This is probably due to the fact that the χ^2 distance and the KLD are heavily influenced by features with low support – an effect that can be observed frequently in our dataset. Thus, the observation model, which is based on the histogram intersection, is given by

$$p(\mathbf{z} \mid c) \propto \sum_{i=1}^{k} \min(\mathbf{h}_i^{\mathbf{z}}, \mathbf{h}_i^{c}). \tag{6.6}$$

The robot can now infer the most likely object class $\hat{c}$ from a single observation by computing

$$\hat{c} = \arg\max_{c} p(c \mid \mathbf{z}) \tag{6.7}$$

using Eq. (6.5) and Eq. (6.6). While this leads to good results on small objects with discriminative features, this strategy is prone to fail on objects that have common partial views. In particular, for the industrial objects depicted in Figure 6.3, tactile images acquired from the bottom of these objects would lead to similar data likelihoods for all industrial objects and thus would not be very informative about the object class.

6.2 Selecting Observation Actions

To uniquely identify an object, the robot therefore has to carry out multiple grasping actions at different height levels. Starting from a uniform prior $p(c)$ over all object classes, we compute the posterior $p(c \mid \mathbf{z}_{1:t})$ according to Eq. (6.5) and Eq. (6.6) after t observations and select the maximum-a-posteriori (MAP) object class according to

$$\hat{c} = \arg\max_{c} p(c \mid \mathbf{z}_{1:t}). \tag{6.8}$$

Intuitively, it seems that an uninformed grasping strategy is not optimal. For example, a large number of grasps might be needed to distinguish different pieces of silverware if they have similar shafts. We therefore propose an informed technique based on concepts from information theory. Our approach seeks to determine the action a that is expected to provide the highest information gain, that is, leads to the highest reduction of uncertainty in the posterior over the object classes. The entropy is defined as

$$H(c) = -\sum_{c \in C} p(c) \log p(c). \tag{6.9}$$

Note that we omit the conditioning on $\mathbf{z}_{1:T}$ here to improve the readability. In our concrete scenario, an action a encodes the (discretized) position at which the robot grasps an object. Let $a_{1:t}$ be the sequence of grasping actions carried out until the current time step t and let $\mathbf{z}_{1:t}$ be the corresponding measurements obtained so far. The robot then has to select the action a_{t+1} that leads to the highest reduction in entropy. Let a be an action under consideration and $\mathbf{z}$ be the corresponding observation that is obtained when carrying out a. The information gain is defined as

$$I(c; \mathbf{z}) = H(c) - H(c \mid \mathbf{z}), \tag{6.10}$$

and is the quantity that the robot seeks to maximize. Since the robot does not know which measurement it will obtain by executing an action a, it needs to integrate over all possible measurement outcomes in order to estimate the information gain, i.e.,

$$E[I(c; \mathbf{z}) \mid a] = \int p(\mathbf{z} \mid a) I(c; \mathbf{z}) \, \mathrm{d}\mathbf{z}. \tag{6.11}$$

In practice, reasoning about all potential observations is intractable since the number of potential measurements grows exponentially in the number of dimensions of the measurement space. A practical approximation, however, is to sum over observations stored in the training set instead of integrating over the whole observation space, i.e.,

$$E[I(c; \mathbf{z}) \mid a] \approx \sum_{\mathbf{z} \in \mathcal{D}_{\text{training}}} p(\mathbf{z} \mid a) I(c; \mathbf{z}). \tag{6.12}$$

The term $p(\mathbf{z} \mid a)$ can be approximated from the training data by counting how often a particular observation in $\mathcal{D}_{\text{training}}$ was obtained when carrying out action a divided by the number of training samples. Depending on the size of the training database, this sum might still be expensive to compute. To further reduce the complexity, one can easily down-sample the training set.

This approach allows us to approximate the posterior efficiently since we can directly utilize the discrete posterior about the identity of an object. The approximations substantially reduce the number of potential observations that have to be estimated by simulation compared to the number of possible measurements the sensor can generate. The ability to carry out such computations efficiently is an important prerequisite for informed action selection.

After having computed the information gain for each action under consideration, we select the action $\hat{a}_{t+1}$ with the highest expected utility

$$\hat{a}_{t+1} = \arg\max_{a} E[I(c; \mathbf{z}) \mid a]. \tag{6.13}$$

Every time the robot has to make the decision of where to grasp next, it uses Eq. (6.13) to determine the action $\hat{a}_{t+1}$ with the highest expected information gain and executes it. As soon as no action provides an expected reduction of uncertainty or the robot reaches a given level of certainty, the identification task is completed.

In addition to the expected reduction of the entropy, one typically has to consider a second quantity when selecting the next action. This quantity is the actual cost of carrying out an action, which needs to be traded off with the expected information gain. In our setting, however, the cost measured in terms of time needed to carry out an action can be assumed to be identical for all actions since the movements of the manipulator are carried out quickly without major differences in the duration. Thus, we ignore the time needed by the manipulator for changing the position. Considering such a cost, however, can be done in a straightforward manner by replacing the information gain in Eq. (6.13) with a suitable *utility* function.

6.3 Experiments

The first goal of our experiments was to find a suitable number of tactile features k of the generated vocabulary and a reasonable choice for the mixing factor α weighting the influence of tactile versus haptic information. With these parameters, we evaluated the recognition rates of different sets of objects. Finally, we used the information-theoretic approach to select the observation actions and evaluated to what extend this improved the recognition rate and recognition convergence.

Table 6.1 Influence of the number of tactile features k on the recognition rate.

number of clusters k	10	20	30	40	50
recognition rate	58.2 %	72.8 %	71.7 %	84.4 %	83.0 %

Table 6.2 Influence of the weight parameter α on the recognition rate.

weighting factor α	0.00	0.25	0.50	0.75	1.00
recognition rate	66.9 %	84.3 %	81.0 %	78.3 %	76.0 %

Tactile Data

For testing our approach, we recorded tactile data for 16 different objects as shown in Figure 6.3. The first 5 objects are industrial parts with a relatively similar shape (for example, a metal cube, a cylinder, and a triangle), while the latter 11 objects were household objects such as cups, toys, and bottles. We created a database of tactile observations by grasping each object on a predefined path (from bottom to top). Some objects were included twice in the dataset, both under 0° and 90° rotation. We obtained a set of $|\mathcal{D}| = 830$ tactile observations for 21 class labels. All experiments were then carried out on this dataset using randomized, 2-fold cross-validation. For each run, we thus had 415 samples for training the model, and 415 (independent) samples for the evaluation.

6.3.1 Vocabulary and Codebook Creation

Before each run of our experiments, a vocabulary was created from the training data $\mathcal{D}$ by running the k-means algorithm. An example of the resulting centroids is given in Figure 6.5. We tried different choices for k and found, by evaluating the recognition rates, that $k = 50$ was a reasonable choice for the number of clusters given our set of objects (see Table 6.1). Alternatively, one could try to find k automatically, for example, by using the Bayes information criterion (BIC). Further, we studied the influence of the weighting factor α in the distance metric of Eq. (6.3). The results are given in Table 6.2. As a result of this preliminary experiment, we chose $\alpha = 0.5$ which means that both the tactile images and the haptically determined object size were considered being equally important. With this, we trained the codebook H according to Eq. (6.4) from the set of labeled training data $\mathcal{D}$.

6.3.2 Recognition Rates

For measuring the recognition rate, we repeatedly chose a random object, and selected $T = 10$ random grasp observations $\mathbf{z}_{1:T}$ of that object from the test set. With this, we obtained a recognition rate of 84.6 % over all 21

Fig. 6.5 Tactile vocabulary created using unsupervised clustering with $k = 50$. Only the left finger of the cluster centroids are depicted.

object classes. In experiments on specific subsets of objects, we found that the household objects among each other were hardly ever confused (96.2 %), in contrast to the industry objects (58.0 %), that look (partially) very similar. The confusion matrices of these experiment are depicted in Figure 6.6. In our dataset, we have also included a tennis ball and a soft ball, two objects that have a very similar visual appearance. In our experiments, we found that these two objects could be separated easily from one another with a recognition rate of 93.8 %.

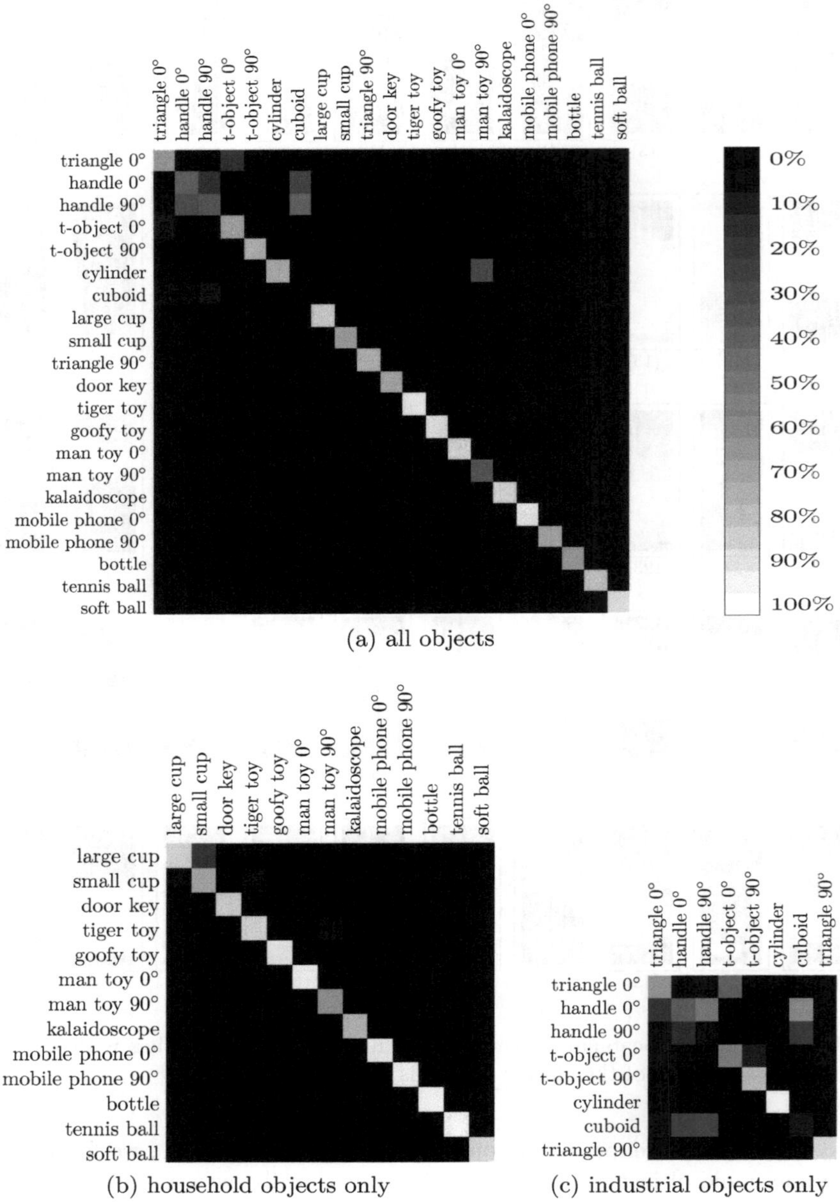

Fig. 6.6 Confusion matrices of the object recognition experiment after 500 object recognition trials with 10 test grasps each on different subsets of our test objects.

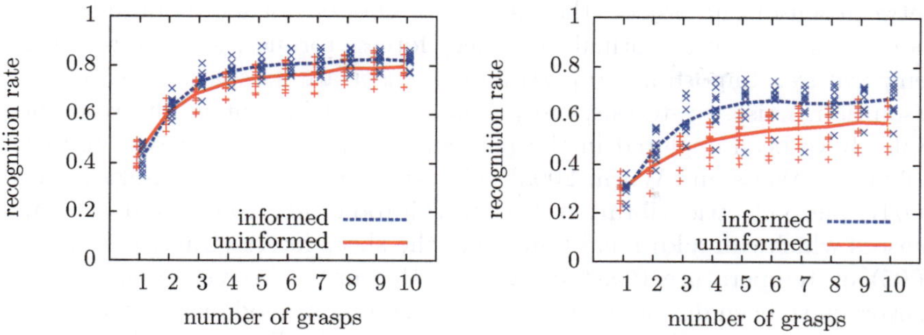

(a) all objects (b) industrial objects only

Fig. 6.7 Comparison of the uninformed and the informed grasping strategy depending on the number of grasp actions.

6.3.3 Active Perception

We also verified whether objects can be recognized with fewer grasps when the robot selects the grasping height based on the expected information gain. We evaluated the recognition rates after each test grasp in 10 independent runs using both the uninformed and the informed grasping strategy. The results are summarized in Figure 6.7. On our full dataset (household and industrial objects), using the information gain strategy performs on average 5.0 % better than random grasping. In particular, one would expect that a better grasping strategy improves the recognition rates on the more difficult dataset of industrial objects. Indeed, we measured a performance gain of 18.9 % of the informed strategy over the random one, raising the average recognition rate from 58.0 % to 67.5 %. In both experiments, a paired t-test showed significantly improved recognition rates when choosing the position that maximizes the expected information gain.

With these experiments, we demonstrated that a robot equipped with tactile sensors can reliably classify a large set of objects using our approach. In particular, we achieved a correct classification rate of 84.6 % over 21 different objects, including objects that are hard to distinguish visually (for example, soft vs. hard tennis ball) and objects that share common parts (such as the industrial objects). Furthermore, we showed that the recognition rate as well as the convergence of the recognition rate improved significantly when the robot used an active grasping strategy. Based on these results, we conclude that mobile manipulation robots can use tactile sensors to recognize the grasped object.

6.4 Related Work and Discussion

Lee and Nicholls (1999) define a tactile sensor as "a device that can measure a given property of an object or contact event through physical contact between sensor and object" that is able to sense one or more of the following modalities: pressure, normal and sheer forces, torques, slip, vibrations, or temperature. Important properties of a sensor are its spatial and temporal resolution, noise, hysteresis, creep, and aging. Different mechatronic principles have been explored in the past such as pressure-sensitive conductive polymers (Weiss and Wörn, 2005), piezo-resistive sensors (Hasegawa et al., 2004), piezo-electric vibration sensors (Motoo et al., 2007), and capacitive sensors which can additionally measure the sheer forces (Chuang and Chen, 2005) or temperature (Castelli, 2002). Recently, also vision-based tactile sensors have been developed (Ueda et al., 2005). A good overview of current developments in tactile sensor hardware is given by Dahiya et al. (2010).

Allen (1988) described an approach to map the 3D shape of objects using tactile sensors. Okamura and Cutkosky (1999) detected ridges and bumps in the material by sliding the robotic finger over an object. Omata et al. (2004) showed that piezo-electric tactile sensors can be used to estimate the deformation properties of biomaterials. Doulgeri and Arimoto (2007) enabled a manipulator to keep continuously physical contact with a surface using force-sensitive fingers, and Bierbaum et al. (2009) presented an approach to explore objects tactically using a five-fingered hand based on potential fields. Petrovskaya et al. (2006) showed how tactile sensors can be used to estimate the 3D pose of objects with known shapes and recently presented an algorithm that globally localizes an object in 6D under performance guarantees (Petrovskaya et al., 2010). As contact sensors only reveal very little information about the probed object, the distributions after the first grasps are highly multi-modal. Therefore, both a good representation of the uncertainty and an effective probing strategy is required. This is a problem with which we also deal in our work.

Several authors addressed the specific problem of recognizing objects using tactile sensors. Russell (2000) proposed to extract geometric features such as point, line, or area contacts from pressure images. Based on these geometric features, the system classified objects into generic classes such as boxes, spheres, and cylinders using hand-coded decision trees. In contrast to his work, our method is not restricted to pre-defined geometric shapes but learns both its own set of tactile vocabulary and feature histograms. Takamuku et al. (2008) used an anthropomorphic hand for recognizing objects. They found that the recognition rate improves significantly after repeatedly opening and closing the hand around the object until the object converges into a discriminative position. More recently, Gorges et al. (2010) recognized different objects based on tactile and kinesthetic sensors using self-organizing maps (SOMs). In their experiments, Gorges et al. found that additional passive degrees of freedom between the robotic fingers and the tactile sensor array helped the objects to settle in characteristic positions, which significantly improved the recognition rates. Pezzementi et al. (2011) recently developed an approach to create large tactile maps of objects by image stitching, and then used these maps to recognize and localize objects from tactile observations using histogram and particle filters.

Tactile sensing requires to deal with actions and observations at the same time, as no tactile observations can be made without touching an object. In their seminal work, O'Regan and Noë (2001) proposed the view that the human brain represents the external world as sensorimotor experiences. Jeannerod (2007) extended this idea towards an object-oriented view, i.e., that objects have both perceptual and motor-function attributes that should be modeled jointly. Both approaches are strongly related to the concept of object affordances as introduced by Gibson (1977). Both Takamuku et al. (2008) and Gorges et al. (2010) exploited in their work that the shape of an object induces a few stable, characteristic grasps which helps to distinguish objects. In our approach, we explicitly exploit the relationship between grasping actions and tactile observations and achieve a significant improvement of the recognition rate.

In contrast to all previous works on tactile object recognition, our approach ensures that the tactile features match the data as the robot generates them from real data by creating a tactile vocabulary. Also, the learned codebook is well grounded as the robot learns the feature histograms from its own tactile observations. Furthermore, most related approaches have only been evaluated on a small set of different objects while we validated our approach on a large set of 21 different object classes.

6.5 Summary

In this chapter, we presented a novel approach for object recognition using tactile observations based on the bag-of-features model. The robot creates a feature vocabulary for the tactile observations based on unsupervised k-means clustering and learns a codebook over this vocabulary to describe the appearance of objects. Subsequently, it can use the learned codebook to recognize a large set of objects. To reduce the number of required grasps, we extend our approach to active grasping with a decision-theoretic framework that uses the expected information gain of potential grasp actions. In experiments on a real manipulation robot, we validated our approach on a large set of real-world objects and achieved accurate recognition results. Furthermore, we demonstrated that our approach on active grasping significantly improves the recognition rate. As a result, our approach enables manipulation robots to quickly identify the grasped objects.

Chapter 7
Object State Estimation Using Tactile Sensors

In the previous chapter, we introduced tactile sensors as an additional source of information that enables a manipulation robot to identify the grasped object. To recognize the object, the robot acquired *static* tactile images after grasping was complete. In contrast to this, we focus in this chapter on features that describe the *dynamics* in the tactile response while the robot is grasping or manipulating an object. As we will show, the dynamic components of the tactile signal can be used to infer several aspects of the internal state of an object. For example, these features allow a robot to detect whether a grasped bottle contains liquid and whether its cap has been properly closed. This information is highly relevant for domestic robots that fulfill service tasks such as tidying up.

We motivate this problem with an example. The robot depicted in Figure 7.1a is given the task of tidying up a table full of bottles. To achieve this task, the robot has to decide which bottles are full or empty and, therefore, can be disposed or have to be placed in the fridge. As the bottles are opaque, it is difficult to determine their fill state purely from vision. The approach presented in this chapter enables a robot to estimate this state by manipulating the object, i.e., by squeezing and rolling it as shown in Figure 7.1b.

In this chapter, we develop a novel set of dynamic tactile features that most manipulation robots equipped with force-sensitive finger tips can extract while they are grasping an object. Our set includes features that describe the *deformation properties* of an object such as the average *compression ratio* and *compression velocity*. Based on these features, we learn a decision tree classifier and show that a robot can use it to reliably estimate both the identity of an object and its internal state. As an application, we demonstrate that a robot can distinguish open from closed containers and full ones from empty ones. To show that this is a hard perception problem, we conduct a comparative study with human test subjects that perform the same discrimination task. In further experiments carried out on a real robot, we demonstrate that a robot can detect the liquid in a container by slowly rolling it.

In Section 7.1, we develop our approach based on a set of dynamic tactile features. We evaluate our algorithm on a large set of real data and compare the performance of the algorithm with that of humans in a similar discrimination task in Section 7.2. Inspired by the feedback obtained in this human

J. Sturm: *Approaches to Probabilistic Model Learning*, STAR 89, pp. 141–160.
DOI: 10.1007/978-3-642-37160-8_7

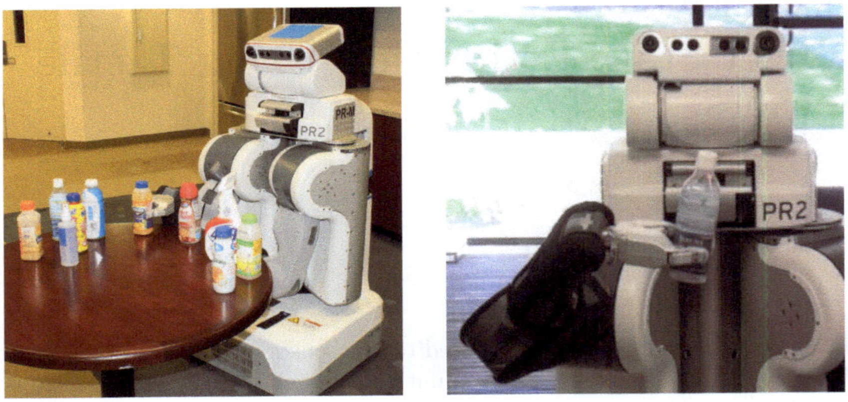

(a) (b)

Fig. 7.1 (a) To clean the table, the service robot needs to decide which bottles it can dispose because they are empty. (b) The robot detects liquid in a bottle by actuating it.

study, we introduce in Section 7.3 an additional dynamic feature based on the high-frequency components in the tactile sensor signal. In a second set of experiments, we demonstrate that this feature can be used to detect liquid in a container. Finally, we conclude this chapter with a discussion of related work in Section 7.4.

7.1 Generic Tactile Features for State Estimation

We assume that a mobile manipulator has a force-sensitive finger tips installed in its gripper that reports at each point in time its position $p(t) \in \mathbb{R}$, velocity $\dot{p}(t) \in \mathbb{R}$ and the force $f(t) \in \mathbb{R}$ acting on its fingers. In this section, we formulate our approach based on these raw (and noisy) sensor measurements. Furthermore, we assume the existence of a controller that we can use to apply a specified force on an object. The aim here is that the controller should not damage the objects but still should be able to grasp the object firmly.

7.1.1 Feature Extraction

In this work, we concentrate on internal state estimation using two-fingered grasps. Such grasps involve pinching an object between its fingers. In our implementation, we used a robotic gripper with two fingers (see Figure 7.2), in principle, our approach also applies to multi-fingered hands.

In preliminary studies, we found that the position, velocity, and force profile of a prototypical grasp has a shape schematically depicted in Figure 7.3. The distance $p(t)$ represents in our case the distance between the two fingers,

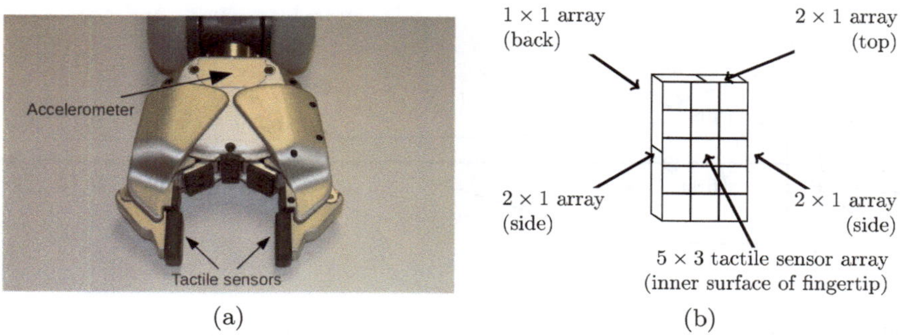

Fig. 7.2 (a) Image of the PR2 gripper used in our experiments. (b) Schematic drawing of the tactile sensor consisting of 22 cells installed in both finger tips.

but could also refer to the volume of the enclosed region of a multi-fingered hand. It decreases until contact with the object is made. The object may deform slightly but ultimately will result in a steady state where the distance between two fingers stays constant. $\dot{p}(t)$ corresponds to the velocity of the fingers. $f(t)$ represents the total force measured at the fingertips using the tactile sensors. Before contact is made, this value is zero. The spike indicates the onset of contact. After the impact, the force reduces again as the object gets deformed and the fingers decelerate. After a while, the motion of the fingers stops and a steady state is reached.

From these profiles, we identified two important points in time: the moment the gripper makes first contact with the object t_{first} and the time t_{steady} after which the sensor values have converged to a steady state. In practice, we require for the first contact detection that both fingers are in contact with the object, i.e., that the force measurement of both fingers is above a threshold F. t_{steady} denotes the point in time where the gripper comes to rest, i.e., its velocity drops below a value V:

$$t_{\text{first}} = \arg\min_{t} |f(t)| > F, \quad (7.1)$$

$$t_{\text{steady}} = \arg\min_{t>t_{\text{first}}} |\dot{p}(t)| < V. \quad (7.2)$$

At moment t_{first}, we extract the first contact distance $p_{\text{first}} = p(t_{\text{first}})$. This is the distance between the two fingertips when contact with the object is first achieved. Note that this is a measure of the uncompressed size of the object. The second feature is the distance between the two fingertips after the gripper has compressed the object fully. We label this the steady state distance

$$p_{\text{steady}} = p(t_{\text{steady}}). \quad (7.3)$$

Note that this distance is a function of both the material and geometric properties of the object and of the internal state of the object, i.e. whether the object is open or closed and full or empty.

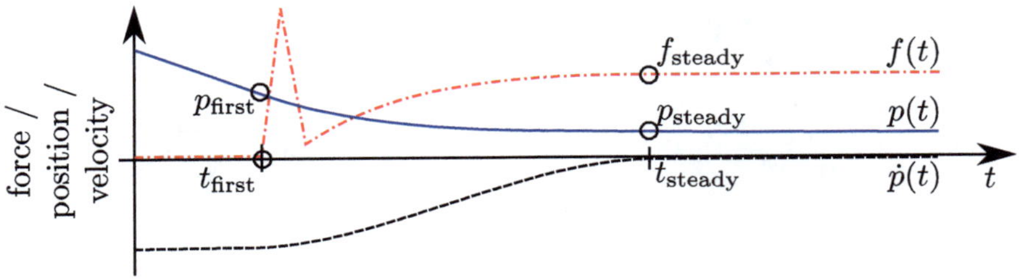

Fig. 7.3 Illustration of a generic force, position, and velocity profile while grasping an object.

Table 7.1 Proposed set of tactile features.

Feature	Description
p_{first}	finger distance at first contact
p_{steady}	distance after which grasping is complete
f_{steady}	force sensed after grasping has completed
Δt	time between first contact and steady state
$\Delta p/\Delta t$	average compression velocity
$\Delta f/\Delta t$	average compression ratio

Another useful feature is the time that it takes between making contact with the object and coming to a rest, denoted by

$$\Delta t = t_{\text{steady}} - t_{\text{first}}. \tag{7.4}$$

Additional features are defined using the force measured by the fingertip sensor array. Let f_{first} be the measured force when both fingertips make first contact with the object and f_{steady} be the measured force once the finger have come to rest. Two other useful features are the average velocity $\Delta p/\Delta t$ of compression and the average rate of change of the fingertip center sensor force $\Delta f/\Delta t$, which can be computed from the features from above as follows:

$$\Delta p/\Delta t = (p_{\text{steady}} - p_{\text{first}})/\Delta t, \tag{7.5}$$

$$\Delta f/\Delta t = (f_{\text{steady}} - f_{\text{first}})/\Delta t. \tag{7.6}$$

The average velocity $\Delta p/\Delta t$ represents the rate at which the object gets compressed and can differ based on the material properties and the geometry of the object. Equivalently, $\Delta f/\Delta t$ could be thought of as representing an average compression ratio. For computing the measured force, we sum over the measured forces of all cells in the tactile sensor array.

These six generic features can be easily extracted by most robots equipped with tactile sensors while grasping an object. Although we do not claim that this list is complete, we were able to reliably estimate the internal state of various containers.

7.1.2 Decision Tree Classifier

Using the tactile features defined above, we gathered data for a large number of different objects. For each trial, we obtained measurements for the 6-dimensional feature vector $\mathbf{a} \in \mathbb{R}^6$, i.e.,

$$\mathbf{a} = (p_{\text{first}}, p_{\text{steady}}, f_{\text{steady}}, \Delta t, \Delta p/\Delta t, \Delta f/\Delta t)^T, \tag{7.7}$$

and a label $c \in C$ describing the object's class and internal state. As a result, we obtained a training database $\mathcal{D}$ containing a sequence of attribute-class tuples $(\mathbf{a}, c)$. A summary of our six generic tactile features is given in Table 7.1.

Subsequently, we applied a C4.5 decision tree classifier on our training data as introduced in Chapter 2. We also tried other supervised classifiers such as support vector machines and neural networks, from which we obtained similar (or slightly worse) results. The reason for this might be that all algorithms are able to extract almost the same amount of data from the training set. The advantage of decision trees over other classifiers is that the learned concepts can intuitively be interpreted. The C4.5 decision tree classifier (Quinlan, 1992) is an extension of the ID3 algorithm that can additionally deal with continuous attributes.

7.1.3 Experiments

The hardware used for the experiments described in this chapter is part of the PR2 robot from Willow Garage. The PR2 is a general-purpose mobile manipulation robot with two arms. Each gripper (see Figure 7.2a) has one degree of freedom which is actuated by a brushless DC motor with a planetary gearbox and an encoder. The rotary motion of the motor is converted into linear motion of the two fingertips of each gripper. Thus, the PR2 gripper is essentially a parallel jaw gripper with 1-DOF. We used the encoder for measuring the finger position $p(t)$ and velocity $\dot{p}(t)$. The gripper can apply a maximum force of 200 N but is software limited to 100 N. Note that this is also approximately the amount of force that a human can apply by pinching the forefinger and thumb together.

Tactile Sensor

Each finger has a capacitive sensor consisting of 22 individual cells mounted on the fingertip. A 5×3 array is mounted on the parallel gripping surface itself while 2 sensors are mounted on the tip of the fingertip, 2 sensors on each side of the fingertip and one on the back, see Figure 7.2b. For this set of experiments, the data from the inner surface of each fingertip was fused into a single force measurement $f_{\text{raw}}(t)$ by summing over all sensor cells. The sensors are capacitive-based pressure sensors and respond to normal pressure exerted on the fingertips. We recorded a calibration curve $g(f_{\text{raw}}) = f_{\text{calibrated}}$ for the sensors using a load cell. The calibration curve as depicted

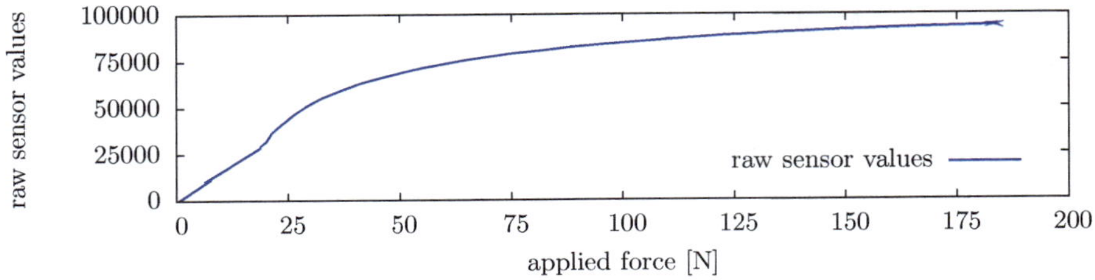

Fig. 7.4 Calibration data relating raw sensor values to forces calibrated using a load cell.

in Figure 7.4 was used as a lookup table. As a result, we obtain calibrated sensor values $f(t) = g(f_{\text{raw}}(t))$ measured in Newtons. Measurements from the tactile sensors on the grippers are obtained at 25 Hz while proprioceptive joint data is measured at 1 KHz. All the joints on the robot are torque controlled in a 1 KHz soft realtime loop. An accelerometer in the gripper measures accelerations in the frame of the gripper which is sampled at 3 KHz.

Switching Velocity-Force Controller

We explored different controllers for the gripper to achieve the objective of grasping objects without crushing them. A pure velocity controller $c_{\text{velocity}}(\dot{p}(t), t)$ makes the gripper approach an object slowly, but after it contacts the object, it increases its force output in order to establish a constant velocity $\dot{p}_{\text{target}}$ and thereby crushes the object. Another option is to use a force controller $c_{\text{force}}(f(t), t)$. Such a controller can hold an object in the hand firmly, by trying to apply a constant force f_{target}. With a constant force controller, the gripper continuously accelerates until contact is achieved. This can lead to high velocities at impact. As an example, see Figure 7.5, where the gripper was grasping a very rigid object (here, a wooden block). The significant impact force applied to the object on contact can easily damage rigid, but delicate objects, like eggs. Of course, the applied constant force could be reduced to deal with such cases. In practice, however, if the commanded force is below the force required to overcome static friction, the gripper does not move at all.

Driven by these considerations, we chose to create a switching controller: first, we close the gripper slowly around an object using the velocity controller until it makes contact with the object. Then, we switch seamlessly to the force controller in order to gently measure the object's deformability properties, i.e.,

$$c_{\text{grasping}}(t) = \begin{cases} c_{\text{velocity}}(\dot{p}(t), t) & \text{while } f(t) = 0 \\ c_{\text{force}}(f(t), t) & \text{thereafter.} \end{cases} \tag{7.8}$$

This switching controller has two parameters: both the initial velocity $\dot{p}_{\text{target}}$ and the probing force f_{target} have influence on the executed grasp.

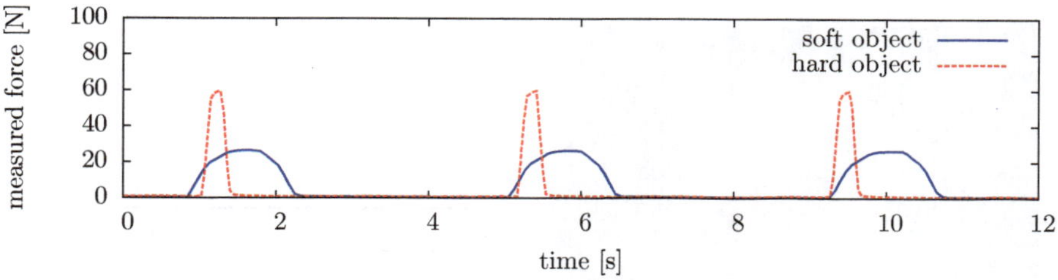

Fig. 7.5 Measured net fingertip force (N) for grasping a wooden block (hard) and a rubber toy (soft) when using a pure force controller. The high impact forces can destroy delicate, but rigid objects, like eggs.

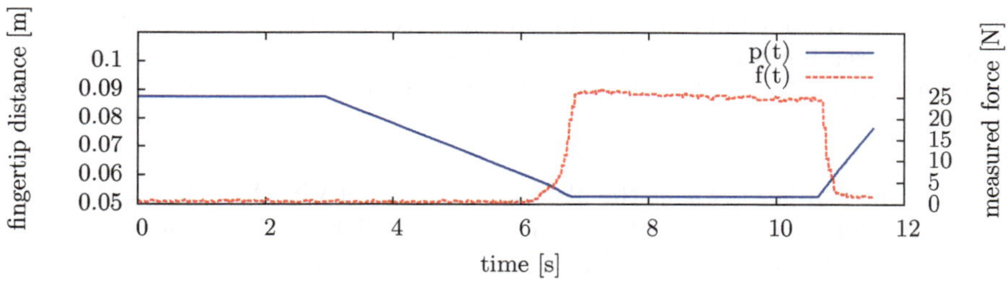

Fig. 7.6 This plot shows the reduced impact forces when using our switching controller. Note that the fingertip force does not spike above the desired probing force on impact.

The result of the switching velocity-force controller can be seen in Figure 7.6. Here, a wooden block was grasped by the gripper using the new controller. The peak force acting on the object is significantly lower. In preliminary experiments, we found that this controller was successful in grasping eggs without crushing them.

Evaluation of Object Recognition and State Estimation

In this section, we describe our experimental setup and subsequently present our results on the recognition rate of object classes and internal states using our approach.

The container classes present in our training set are Odwalla fruit juice bottles, Naked fruit juice bottles, soda cans and plastic water bottles, see Figure 7.7a. The internal states of these containers are: closed and full, open and full, open and empty, and closed and empty (except for the soda cans, which cannot be closed again after having been opened).

We started the acquisition of training samples with the gripper fully open. The containers were placed one at a time between the gripper fingertips, i.e., we did not deal with localizing the object prior to grasping via vision nor with

(a)

	a	b	c	d
Odwalla bottles = a	58	1	0	1
Naked bottles = b	8	40	0	0
Softdrink cans = c	0	0	41	3
Water bottles = d	0	0	1	76

(b)

Fig. 7.7 (a) Bottles and cans used in our experiments. From left to right: Odwalla bottles, water bottles, Naked bottles, Softdrink cans. (b) Confusion matrix for recognizing the class of the container. The recognition rate is 93.9 %.

moving the gripper towards the object. The gripper was then closed using the switching force velocity controller described earlier. Once the gripper came fully to rest, the controller waited for a small interval of time before opening the gripper fully. During each trial, the features described in Section 7.1 were extracted and written to a file.

We collected data for each of the internal states for each container class using our controller. We carried out a total of 66 trials with 12 Odwalla fruit juice bottles in 4 different internal states, 80 trials with 16 water bottles in 4 different internal states, 42 trials with 12 cans with 3 different internal states, and 41 trials with 10 Naked fruit juice bottles with 4 different internal states. We used different instances of each container class in collecting the data to account for variations within a container class. We also rotated the containers between taking measurements to account for variations in surface properties of individual containers. All this data was collected with the probing force set to $f_{\text{target}} = 20\,\text{N}$. We also collected a subsequent dataset just for the Odwalla fruit juice bottles using three different probing forces of 17, 20 and 23 N. This involved conducting 24 trials for each internal state for a total of 96 trials for all the 4 internal states for each probing force.

To test our classifier, we used ten-fold cross-validation for each experiment. First, we divided the stratified dataset into 10 parts. Second, we learned the classifier on 9 parts and used it subsequently to classify the test instances from the remaining part. This was repeated for each of the ten folds such that we ended up with target class predictions for all instances in the dataset. Finally, we compared the predictions to the true target class and computed the recognition rate as the ratio between correct and incorrect instances.

In the first experiment, we found a 93.9 % accuracy in recognizing the different liquid containers. Figure 7.7b shows the confusion matrix for this experiment. From the learned decision trees, this high performance can mainly

Table 7.2 The recognition rate depends on the probing force parameter f_{target}. Here, the recognition rates for the Odwalla fruit juice bottles are given.

f_{target}	Recognition Rate
17 N	69.8 %
20 N	83.3 %
23 N	94.8 %

be attributed to the different size of objects, thus p_{first} and p_{steady} are very discriminative for this set of containers. Note that our approach is not meant to compete with other senses like vision, but is meant to complement other approaches and could, for example, be used to confirm a particular object class hypothesis while the robot grasps an object.

After that, we evaluated the recognition rate of the internal state of a container, given its class. We found that the recognition rate strongly depends on the particular container. This result is not surprising, as obviously feeling the internal state of a container strongly depends on how well it manifests its internal state to the outside, i.e., in its tactile appearance. Interestingly, we found that the Odwalla bottles were separable the easiest. Their internal state was estimated correctly at 94.8 %, compared to 74.4 % for cans, 58.3 % for Naked bottles, and only 32.5 % for water bottles. The reason for the low performance on water bottles could be that they are made of very flimsy plastic and tend to deform unpredictably.

We also found that the recognition rate was a function of the parameters of our switching controller. While the influence of the initial grasping velocity $\dot{p}_{\text{target}}$ was negligible, we found that choosing a good probing force f_{target} could improve the recognition substantially (see Table 7.2). This parameter determines how hard the gripper probes into the object and should therefore be carefully selected according to the object class. In the case of the Odwalla bottle, we found, for example, the stronger probing force of $f_{\text{target}} = 23\,\text{N}$ to be more informative than weaker ones, yielding a recognition rate of the internal state of 94.8 %. The confusion matrix for the specific case of recognizing the internal state of an Odwalla bottle is shown in Table 7.3.

In a combined experiment, where we let the robot estimate both the container class and the object internal state except for water bottles (resulting in 11 possible combinations), we obtained a recognition rate of 63.8 %.

By evaluating the learned decision trees, we found that the open and full bottle tends to be compressed for the longest time, i.e., Δt is large. The steady state force f_{steady} differentiates between the open and empty bottle and the empty and closed bottles while the steady state distance p_{steady} differentiates the closed and full bottle very easily. However, when we repeated this experiment with bottles that had been subjected to repeated compressions, the recognition rate decreased again to 81.0 %. This is not surprising

Table 7.3 Confusion matrix of our approach for recognizing the internal state of an Odwalla fruit juice bottle from the tactile appearance using a robotic gripper ($f_{\text{target}} = 23\,\text{N}$). The recognition rate is 94.8 %.

	a	b	c	d
full closed = a	24	0	0	0
empty open = b	0	20	1	3
full open = c	0	0	24	0
empty closed = d	1	2	0	21

considering that the classifier was trained on data from fresh bottles while the testing was now done with bottles that had been subject to repeated stress.

Our experiments show that our approach enabled the robot to recognize the container classes at the high rate of 93.9 %. The performance on recognizing the internal class strongly depends on the object class, ranging from excellent 94.8 % for the Odwalla bottle down to only 32.5 % for the flimsy water bottles.

7.2 Comparative Human Study

We designed a human study to compare the performance of the robot to that of humans for the internal state estimation problem. The aim of the study was to find out if, using only tactile feedback, humans could achieve comparable recognition rates for the task of recognizing the internal state of an object. Figure 7.8a shows the experimental setup used for this study. We asked the test subjects to recognize, using only tactile information from squeezing a bottle, the internal state of the bottle. Beforehand, we provided the test subjects with the opportunity to train until they were confident about their ability to discriminate between the different internal states of the bottles. Then, we asked each test subject to identify the internal state of 16 different bottles sequenced in a random order. The subjects were instructed not to move or pick up the bottles and could not see the bottles while they were grasping them. To simulate the two-fingered grasp used by the gripper, we asked the test subjects to use only their thumb and index finger for the grasping task. Additionally, the test subjects wore noise-canceling headphones to minimize the sound cues that subjects could pick up. In total, 17 persons participated in our study.

Figure 7.8b shows the overall confusion matrix for all the trials together. The average recognition rates for all the subjects was 75.2 %. The highest recognition rate was for bottles that were full and closed. There was considerable confusion between the empty/closed and full/open bottles. Based on a questionnaire filled out by the subjects at the end of the test, we found that most subjects were using features similar to the ones chosen for the learning approach. The two most cited features by the human subjects were the total

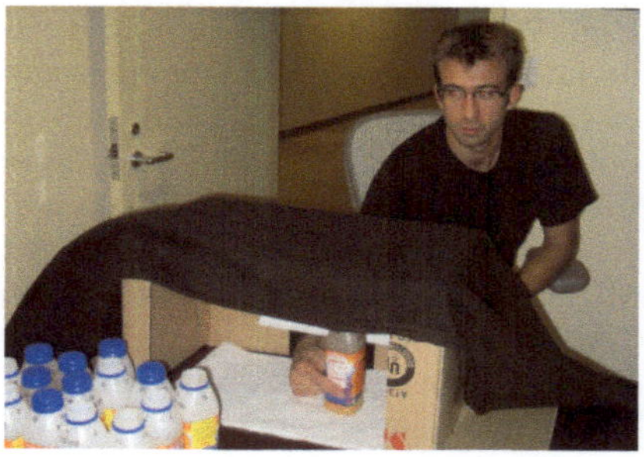

(a)

	a	b	c	d
empty open = a	48	8	5	0
empty closed = b	5	41	1	3
full open = c	16	11	55	2
full closed = d	2	8	7	63

(b)

Fig. 7.8 Experimental setup and results of the comparative human study. (a) The human test subject estimates the state of a juice bottle. (b) Confusion matrix for all human subjects for recognizing internal state of an Odwalla fruit juice bottle. The recognition rate is 75.2 %.

compression distance and the rate at which the bottle returns to its original shape. The second feature is easier for humans to detect than for the robot since the grippers on the robot are not easily back-drivable. The most successful test subjects cited a different set of features to discriminate between the bottles. They used high-frequency feedback from tapping the bottle with their fingers to detect the presence or absence of liquid in the bottle. We took this inspiration to develop a novel tactile feature that enables a robot to extract similar information and to use it for the detection of liquid in containers. We present this extension of our approach in the next section.

7.3 High-Frequency Tactile Feature for State Estimation

Several human subjects cited their use of high-frequency feedback from tapping the container with their fingers as critical to the success of their recognition efforts. Gaining such information with a robot, however, requires the ability to excite an object sufficiently fast and the ability to sense the response of the object to such actuation. Most robotic hands do not have the high bandwidth necessary for such actuation. In our case, the gripper by itself is not fast enough to excite the contents of the container in such a manner. However, we found that we could achieve the desired effect by using the entire arm of the PR2. In this section, we expand on the details of actuation and sensing for the PR2 to be able to use high-frequency information to detect the internal state of objects. We present experimental results that show how this information can prove useful, in particular, in detecting the presence of liquids inside containers.

Fig. 7.9 Containers used in the experiments to determine the presence of liquid. From left to right, top to bottom: Sauve, Nesquik, Dry Erase Cleaner, Zero Calorie, 409 containers, tape dispenser, Odwalla Orange, Might Mango, Summer Lime, Green Tea, dummy weight, Tropicana, water bottle, Coffee Mate, CVS HP and Palmolive containers.

Figure 7.1b show snapshots of the actuation procedure for experiments designed to excite the internal contents of objects. The objects used in the set of experiments are liquid containers. Each container is grasped firmly in the gripper of the PR2 and rolled from side to side at about 0.6 Hz. This motion is designed to excite the internal contents of closed containers. Note that if the object were an open container, its contents would spill out as a result of this motion.

In preliminary experiments, we also tried horizontal movements that would allow for open containers. However, we found the PR2 to be too slow to sufficiently excite the contents of the probed containers. This forced the use of the strongest joint on the robot (the joint that rolls the wrist from side to side) to sufficiently excite the contents of the container by forcing the liquid to slosh around under the influence of gravity. In our belief, the overall approach is more generic and should be executable on any robot that is capable of exciting the internal contents of objects at higher frequencies. In particular, we believe that it is also applicable to open containers with liquid in them if the robot were capable of shaking the container from side to side at a high frequency while maintaining it level.

Experimental results are presented here for the 15 different containers depicted in Figure 7.9. Five trials were carried out for 13 containers with liquid in them. The liquids in the different containers included water, orange juice, mango juice, shampoo and cleaning fluid, thus representing a good range of viscosity and content. Most of the containers were filled to half their volume with liquid. The *Odwalla Orange* container was tested with two amounts of liquid in it - full and half-full, the *Dry Erase Liquid* container was tested with a full volume of liquid and the *Mighty Mango* and *Sauve* containers were tested when about a quarter full. Five trials each were also carried out for 13 of the containers with no liquid in them, i.e., the contents of the container were completely emptied out. An additional five trials were carried out for a rigid weight that weighed about the same as some of the containers with liquid in them. Figure 7.9 shows all the containers used in the experiment.

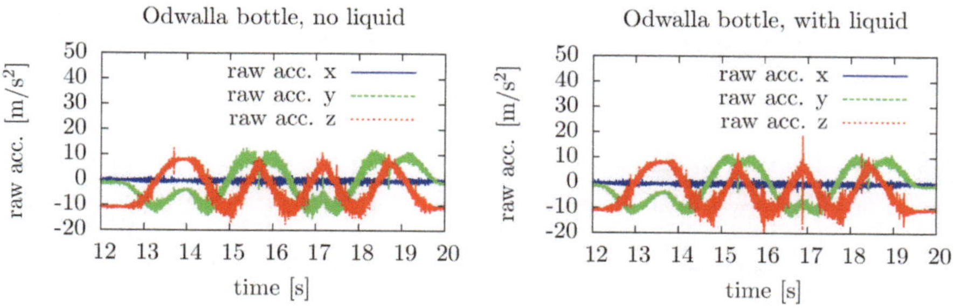

(a) raw measurements of the accelerometer

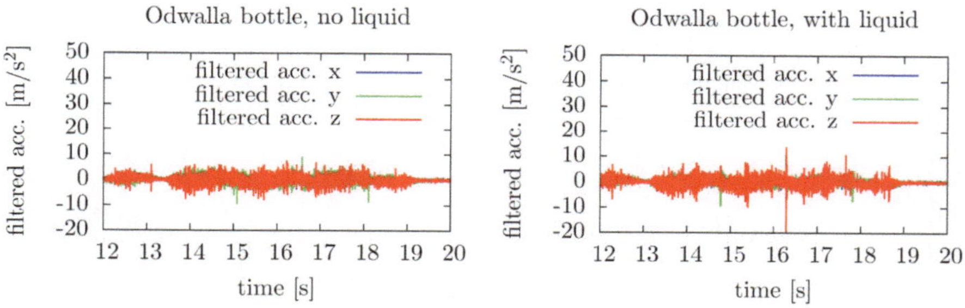

(b) after 5 Hz high-pass filtering of the accelerometer data

Fig. 7.10 Accelerometer data corresponding to a container without liquid (left column) and with liquid (right column) for the Odwalla orange juice bottle.

7.3.1 Training Data

The data measured and recorded for each trial included acceleration data from the accelerometer in the gripper of the robot, tactile sensor data from all 44 elements of the tactile sensors on both fingers of the PR2 gripper and joint positions, velocities, and torques for all the moving joints in the arm of the PR2.

Figure 7.10 and Figure 7.11 represent two example sets of sensor data for the time period when the container is being rolled: the plots on the left of each figure correspond to data for a container with no liquid in it while the plots on the right correspond to data for a container with liquid in it. The accelerometer data in Figure 7.10a is noisy and dominated by the component corresponding to the motion of the container. Figure 7.11a shows the individual tactile sensor responses (for all 44 tactile sensors) over the same period. It is clear that the raw data in this form is not very useful to discern the presence or absence of internal contents in the container.

Our key idea is that liquid sloshing around in a container will produce high-frequency responses that the robot can measure. We preprocessed both

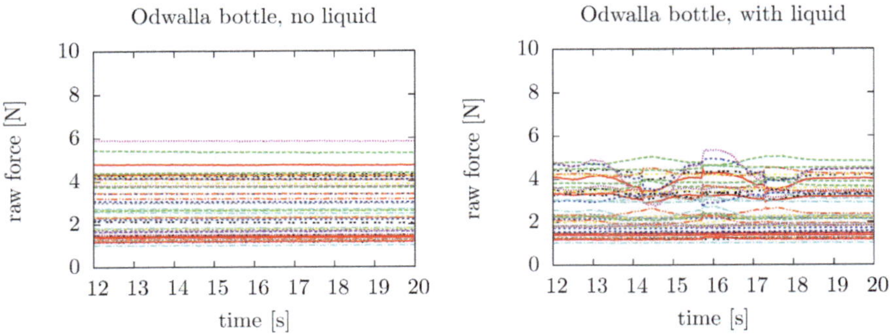

(a) raw measurements of the 44 tactile sensor cells

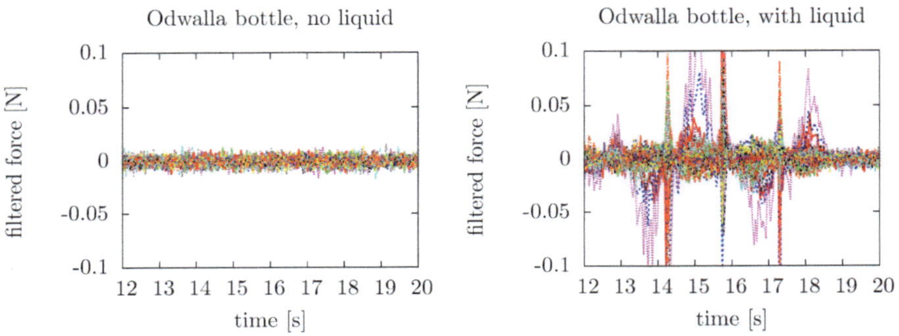

(b) after 5 Hz high-pass filtering of the 44 tactile sensor cells

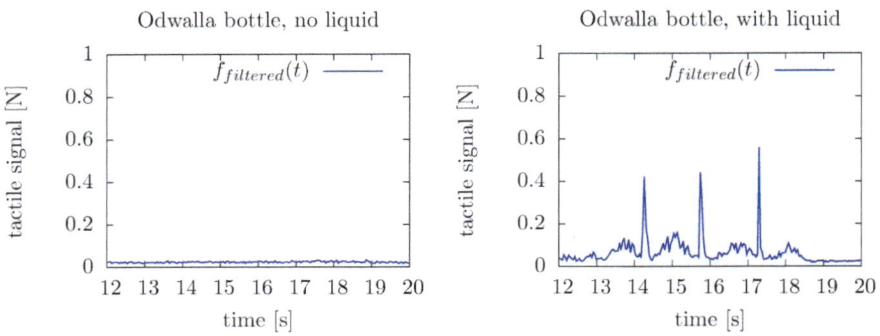

(c) combined tactile feature

Fig. 7.11 Tactile sensor data corresponding to a container without liquid (left column) and with liquid (right column) for the Odwalla orange juice bottle. The sloshing liquid produces very clear spikes in tactile signal (bottom right).

the acceleration data and the tactile data with a high-pass filter. After filtering, we condensed the 3- and 44-dimensional signal for the acceleration and tactile data, respectively, into a single, real-valued signal by computing the Euclidean norm of the signal vector.

Figure 7.10b and Figure 7.11b show the resulting signal from acceleration data and the tactile sensor data filtered through a 5 Hz high-pass filter, respectively. The filter attenuates the low-frequency components corresponding to the rotation of the container. While the accelerometer signal is slightly different for the two cases, the higher-frequency components in the tactile sensor data, however, are clearly different when the container has liquid in it. The sloshing of the liquid in the container due to its excitation during the rolling of the container results in a spike in the tactile sensor pressure whenever the direction of rotation undergoes a change. This information can easily be computed online and is used to train a classifier that can detect the presence of liquids in containers. For that, we compute the variance in the signal measured by the tactile sensor while the object is being actuated (Figure 7.11c). The difference in this value for the two cases (presence or absence of liquids), is large and consistent across different containers. A summary of our results is given in Table 7.4.

In contrast to the tactile sensor, the accelerometer is significantly affected by the motion of the arm. This makes the acceleration data noisy. It is possible that an accelerometer placed closer to the object (for example, on the fingertips) may be able to capture better object information. The tactile sensors on the PR2 are closer to the object and the multiple sensor cells on the sensor can measure the response at multiple points on the object at the same time.

7.3.2 Feature Extraction

The high-pass filters we apply to each of the tactile sensors $i \in 1, \ldots, 44$ are first-order Butterworth filters designed with a cutoff frequency of 5 Hz for the sampling rate of 24 Hz. A similar filter (designed for the sampling rate of 3 KHz) is also applied to the accelerometer signal. The use of the 5 Hz cutoff frequency was motivated by experiments that showed that humans possess tactile receptors that specifically react to signals in the 5–50 Hz range when responding to force disturbances (Johansson and Flanagan, 2009). Let f^i_{filtered} denote the filtered signal for each individual tactile sensor element (here $i \in 1, \ldots, 44$). We combine the signals of all tactile sensor elements into a single signal by computing the Euclidean norm of the filtered signal vectors, i.e.,

$$f_{\text{filtered}}(t) = \left(\sum_{i \in 1,\ldots,44} (f^i_{\text{filtered}})^2 \right)^{1/2} . \tag{7.9}$$

From this combined signal, we estimate the sample mean and variance of this signal as

$$\mu = \frac{1}{n} \sum_{t=1,\ldots,n} f_{\text{filtered}}(t) \text{ and} \tag{7.10}$$

$$\sigma^2 = \frac{1}{n-1} \sum_{t=1,\ldots,n} (f_{\text{filtered}}(t) - \mu)^2, \tag{7.11}$$

where n is the number of data samples while the robot was rolling the object and t refers to the corresponding time indices. In total, we collected data from 136 trials of 15 different containers, see Table 7.4.

For detecting the presence of liquid, we use the estimated signal noise σ as the only tactile feature. The target attribute is binary, i.e., either indicating an empty or a filled container. We train a decision tree classifier and evaluate its performance using ten-fold cross-validation.

7.3.3 Experiments

The learned classifier was able to predict the correct internal state of a bottle correctly in 91.9 % of the cases. Table 7.5 gives the confusion matrix for this experiment.

By looking at the instances for which prediction errors occurred, we found that all five examples of a full CVS HP bottle were incorrectly classified as empty. This bottle is much smaller than the other containers. As a result, the tactile response is relatively small, when compared with the response of heavier containers.

To remedy this problem, we provided in another experiment the weight of the object in the gripper as a second (additional) feature. By using both the weight and the signal noise, we found a 98.5 % correct classification rate. It is important to note that the heavy weighted dummy object was classified correctly as containing no liquid, while the light CVS HP bottle was correctly classified as containing liquid. When looking at the learned decision tree, we found that the resulting classifier uses both the tactile signal and weight for predicting the fill state of a container.

In our experimental setup, one might argue that the weight is a strong indicator of the internal state of an object. While it may have some contribution to the amplitude of the observed high-frequency part of the tactile signal induced by the robot's motors, it is worth noting that the corresponding signal shown in Figure 7.12 for the dummy weight displays virtually no high-frequency component, while the signal of the much lighter CVS HP bottle displays a weak but clear signal. This implies that the presence of liquid in the containers plays a significant role for the observed tactile feature $f_{\text{filtered}}(t)$, but that its magnitude depends on the weight of the liquid content.

However, we also found that the tactile signal corresponding to a *slip* of the object also has a high-frequency component. Heavier objects are most

Table 7.4 Evaluation of the high-frequency tactile feature for various objects.

Object	State	Weight [kg]	Trials	Avg. Tactile Feature [N]
Dummy object	no liquid	0.199	5	0.000 ± 0.000
409	no liquid	0.074	5	0.004 ± 0.001
	with liquid	0.459	5	0.177 ± 0.104
Coffee Mate	no liquid	0.0417	5	0.003 ± 0.002
	with liquid	0.3188	5	0.292 ± 0.333
CSV HP	no liquid	0.0268	5	0.001 ± 0.002
	with liquid	0.160	5	0.002 ± 0.001
Dry Erase Cleaner	with liquid	0.254	5	0.025 ± 0.014
Green Tea	no liquid	0.033	5	0.000 ± 0.000
	with liquid	0.300	5	0.012 ± 0.007
Mighty Mango	no liquid	0.0381	3	0.000 ± 0.000
	with liquid	0.075	5	0.042 ± 0.026
Nesquik	no liquid	0.038	5	0.000 ± 0.000
	with liquid	0.311	5	0.042 ± 0.016
Odwalla Orange	no liquid	0.031	5	0.000 ± 0.000
	half full	0.311	5	0.011 ± 0.009
	full	0.487	5	0.041 ± 0.018
Palmolive	no liquid	0.045	5	0.000 ± 0.000
	with liquid	0.390	5	0.312 ± 0.061
Sauve	with liquid	0.206	5	0.030 ± 0.010
Summer Lime	no liquid	0.029	3	0.008 ± 0.007
	with liquid	0.315	5	0.093 ± 0.043
Tropicana	no liquid	0.032	5	0.000 ± 0.000
	with liquid	0.256	5	0.052 ± 0.048
Water Bottle	no liquid	0.014	5	0.000 ± 0.000
	with liquid	0.253	5	0.179 ± 0.035
Zero Calorie	no liquid	0.042	5	0.000 ± 0.000
	with liquid	0.323	5	0.041 ± 0.022

Table 7.5 Confusion matrix for the recognition rate of the fill state of various objects using high-frequency filtering on tactile data. The recognition rate is 91.9 %.

	a	b
no liquid = a	58	3
with liquid = b	8	67

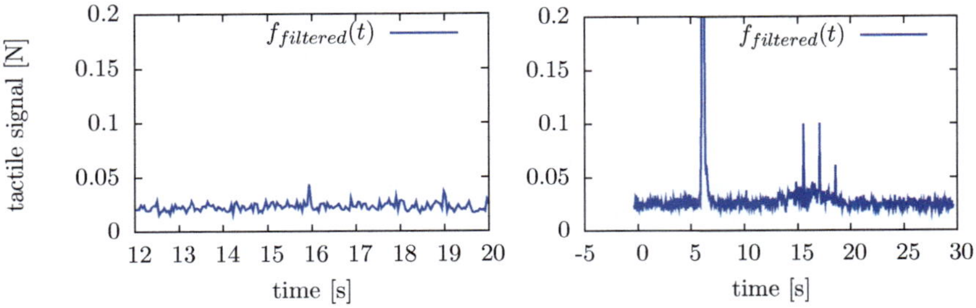

(a) heavy dummy object, no liquid (b) light CVS HP bottle, with liquid

Fig. 7.12 The weight only marginally influences the high-frequency feature. The heavy dummy object in (a) shows virtually no response, while the light CVS HP bottle in (b) produces clear spikes.

likely to slip, especially if they are hard to grasp in the parallel jaw gripper of the PR2. A heavier weight (the tape dispenser in Figure 7.9 weighing about 0.5 kg) does display the same frequency response as containers with liquid as illustrated in Figure 7.12. The ability to detect shearing forces on the fingertips (using a slip sensor) might allow us to separate out the slip component of the signal, but currently, in the absence of such data, our approach is unable to distinguish between the slip of heavy objects that are grasped awkwardly and containers with liquid in them. A stronger gripper that can grasp heavier objects more firmly would also help to reduce the slip. The other possible modification to our approach which may help to reduce the effect of weight is actuating the containers from side to side while keeping them level.

In this section, we demonstrated that the high-frequency components in the tactile signal can be used for estimating whether a container contains liquid. Our approach enabled our manipulation robot to correctly classify objects with 91.9 % accuracy. If the robot additionally used the weight as a second feature, the recognition rate increased to 98.5 %. These results show that tactile sensors can provide valuable information for mobile manipulation robots.

7.4 Related Work and Discussion

In the application scenario introduced in Chapter 1, we motivated our work on tactile sensing with the example of a service robot that has to tidy up objects from a table after a meal. Various gripper designs have been proposed in the past for such tasks. Antropomorphic hands, such as the DLR Hand II (Grebenstein et al., 2011), aimed at emulating the human hand and are usually equipped with a large number of DOFs and a rich sensor suite. Unfortunately, the large number of DOFs makes the analysis of grasp stability highly complex (Bicchi and Kumar, 2001). The alternative are simple but general hands with only very few DOFs. Recent research indicated that such simple grippers with only two- or three-fingers bear an enormous potential for stable and robust grasping. Mason et al. (2012) argued that simpler hands are easier to study and to understand. Inspired by non-robotic pickup tools and prosthetic hooks, they developed various underactuated 1-DOF grippers for bin-picking tasks with three and four fingers, respectively, and devised novel ways to prove grasp stability based on potential fields. Following a similar paradigm, Xu et al. (2009) recently presented a simple robot for domestic object retrieval that they validated on a large set of different objects in various poses. Dollar and Howe (2010) as well as Ciocarlie and Allen (2010) demonstrated that compliance and adaptivity in the gripper can significantly increase the reliability of grasping, in particular when the pose and object identity is not perfectly known beforehand. When the workspace is cluttered, as for example in a cutlery drawer, reliable object segmentation, identification, and localization is not possible before the grasp. In this case, the robot needs to grasp blindly until it successfully singulates and grasps one of the objects. Mason et al. (2012) called this approach "let the fingers fall where they may" and the strategy "grab first, ask questions later". Similarly, Gupta and Sukhatme (2012) and Hausman et al. (2013) used interactive perception to sort a pile of bricks and to pick up objects from a cluttered table, respectively. In these examples, the robot did not know whether a grasp will be successful, and so it had to determine the success of the grasp, the object identity, and the in-hand pose afterwards. To determine this, all of the above approaches used the finger encoder values of the final gripper pose. In contrast to this, we additionally included tactile information and considered the signal dynamics. As a result, our controller was able to firmly grasp an object without crushing it. Furthermore, by using the tactile information, we demonstrated that the robot is able to tell the internal state of the object and could thereby discriminate empty from full bottles.

Several studies from neurophysiology have shown that humans cope very well with modulating the applied grasp force in relation to the expected load force (Johansson, 1996; Williams et al., 2010). Even during dynamic motions such as walking or running, humans always apply the minimum force required to hold an object safely. These coordinative constraints simplify the control by reducing several degrees of freedom during the manipulation tasks. Tactile perception hereby plays an essential role: in experiments with humans, it was shown that the test subjects exerted much more gripping force than needed when their fingertips were anesthetized even if visual feedback was available (Monzee et al., 2003). Furthermore, Johansson and Flanagan (2009) described how humans use the high-frequency components of the tactile signals in pick-and-place tasks.

In their recent survey, Dahiya et al. (2010) reviewed current tactile sensor hardware for manipulation robots, for example the work of Weiss and Wörn (2005) on resistive sensor cells, Ohmura et al. (2006) on a flexible sensor skin,

Ueda et al. (2005) on vision-based tactile finger tip sensors. Both Maeno (2004) and Matuk Herrera (2008) estimated the friction coefficients of the grasped object to avoid slippage. Frank et al. (2010) considered the problem of navigating in environments with deformable objects. By combining a force-torque sensor with a depth camera, they learned the deformation coefficients of various objects. They showed that this information can then be used to minimize the expected deformation costs during trajectory generation. Saal et al. (2010) estimated the viscosity of different liquids by shaking a container. They proposed a strategy to speed up the convergence by actively varying the shaking frequency and amplitude. Another notable approach used acoustic sensors: Sinapov et al. (2009) demonstrated that a robot can discriminate containers from non-containers based on the sound an object produces while the robot performs grasping, shaking, and dropping behaviors. In their later work, Sinapov and Stoytchev (2010) investigated to what respect the information from multiple sensor modalities is redundant. Their analysis included auditory, proprioceptive, and tactile observations. In contrast to these previous approaches, our goal of estimation is different, i.e., we apply tactile sensors to learn a classifier for the internal state of various containers.

7.5 Summary

In this chapter, we presented a novel approach to determine the internal state of objects based on a small set of tactile features. These features can be extracted from the sensor while the robot grasps or manipulates the object. To estimate the internal state, we train a decision tree classifier on a limited set of labeled training data. In experiments carried out on real robots, we demonstrated that a robot using our approach can reliably classify the internal state of the grasped object. Furthermore, we found in a comparative study on human test subjects that the recognition capability of the robot match is similar to the performance of humans. To conclude, our approach enables a robot to reliably detect empty bottles when cleaning up a table which we consider an important ability for service robots operating in domestic environments.

Chapter 8
Learning Manipulation Tasks by Demonstration

To accomplish a particular manipulation task, a robot needs a detailed description of how to execute it. However, it is not possible to specify all potential tasks of a manipulation robot beforehand. For example, robotic assistants operating in industrial contexts are frequently faced with changes in the production process. As a consequence, novel manipulation skills become relevant on a regular basis. For this reason, there is a need for solutions that enable normal users to quickly and intuitively teach new manipulation skills to a robot.

Promising approaches to this problem have been presented in the *imitation learning* community (Argall et al., 2009; Billard et al., 2008; Abbeel et al., 2007; Ijspeert et al., 2002; Bakker and Kuniyoshi, 1996). The key idea behind these approaches is that the robot learns a new manipulation skills by observing a human demonstrator. As a motivating example, consider Figure 8.1. The human instructor illustrates how to clean a white board while the robot observes his motions using visual motion capture. From this presentation, the robot infers the underlying task description. By carefully evaluating the differences and similarities between multiple demonstrations, the robot generalizes the task model so that it becomes applicable to novel situations. This enables the robot to robustly reproduce a learned skill even under varied conditions, for example, to also clean white boards of different sizes.

In this chapter, we consider the problem of learning generalized descriptions of object manipulation tasks from human demonstrations. We employ dynamic Bayesian networks (DBN) as a compact representation where special nodes encode geometrical constraints between the relevant objects in the scene and the hand of the demonstrator. This formulation allows the robot to learn generalized task descriptions from multiple demonstrations so that it can reproduce them also under changed conditions. Furthermore, novel constraints can easily be added during the reproduction of a task. This is, for example, relevant to allow a robot to deal with obstacles. To reproduce a task, the robot searches for the action sequence that maximizes the data likelihood in the DBN. In experiments carried out in simulation as well as on a real manipulation robot, we show that our approach enables robots to efficiently learn novel manipulation skills from human demonstrations and to robustly reproduce them in different situations.

J. Sturm: *Approaches to Probabilistic Model Learning*, STAR 89, pp. 161–178.
DOI: 10.1007/978-3-642-37160-8_8

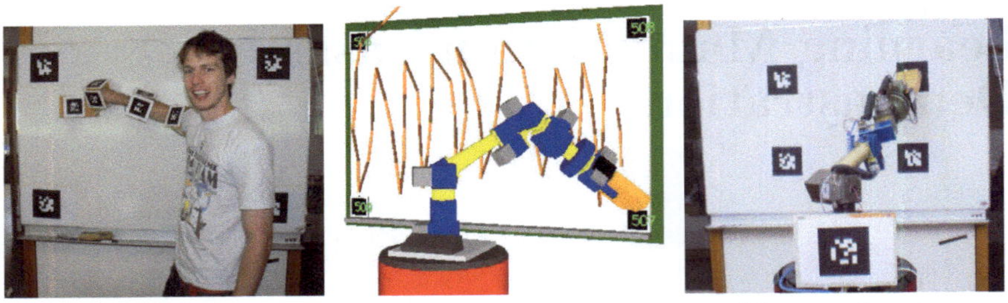

(a) human instructor (b) reproduction (c) generalization

Fig. 8.1 A manipulation robot learns how to clean a white board by observing a human instructor.

The remainder of the chapter is organized as follows: in Section 8.1, we introduce our model for manipulation tasks based on DBNs. In Section 8.1.1, we show how a robot can infer the relevant constraints from motion capture data and describe in Section 8.1.2 how a robot can reproduce a learned manipulation skill in novel situations. We evaluated our approach both in simulation and on a real 6-DOF manipulator, and present our results in Section 8.2. Finally, we conclude this chapter with a discussion of related work in Section 8.3.

8.1 Modeling Manipulation Tasks

We model a manipulation task as a stochastic process that we represent by means of a dynamic Bayesian network as depicted in Figure 8.2. During the execution of a task, the demonstrator $\mathbf{q}$ manipulates the poses $\mathbf{x}$ of objects in the scene. The robot observes the body configuration $\tilde{\mathbf{y}}$ (in joint angles) of the demonstrator and the poses of objects in the scene $\mathbf{y}$ (in Cartesian coordinates). From that, the robot learns the task constraints both in configuration space $\tilde{\mathbf{r}}$ and Cartesian space $\mathbf{r}$. During reproduction, the robot uses these constraints to infer the next most likely body configuration which it subsequently executes.

Concretely, our goal is to estimate the constraints $\mathbf{r}^t$ between n objects in the world and the manipulator as well as constraints $\tilde{\mathbf{r}}^t$ in the body configuration from the demonstrations of the human teacher. These constraints encode the essence of the manipulation task. We learn the object constraints $\mathbf{r}^t \in \mathbb{R}^{3p}$ between p objects that we encode as the 3D position of each object with respect to the position of the end effector. Further, we learn the configuration constraints $\tilde{\mathbf{r}}^t \in \mathbb{R}^d$ that correspond to the joint angles of the d-DOF manipulator (or demonstrator). We model each of these constraints as a Gaussian distribution that we estimate for each time step t from the human demonstrations. The general idea is that relevant constraints will lead

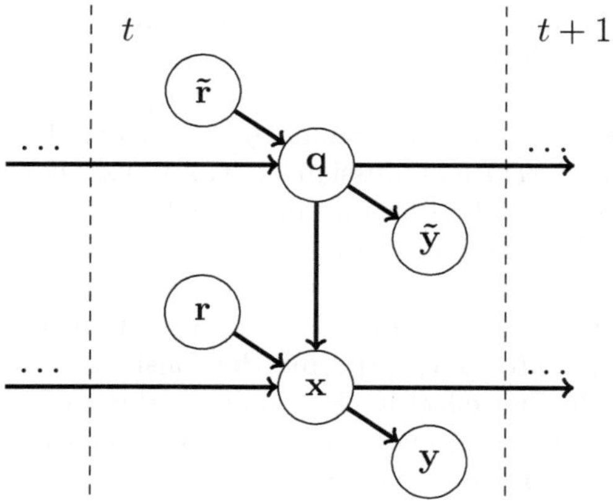

Fig. 8.2 We model the imitation learning problem using a dynamic Bayesian network (DBN).

to peaked probability distributions, while irrelevant constraints will manifest themselves as distributions with high variance.

During task reproduction, we use the learned constraints to infer a sequence of consistent configurations that maximizes the likelihood of the learned task model. At this time, the robot uses the learned probability distributions encoding the task constraints to generate an action sequence that reproduces the task given the current world state.

For simplicity, let us consider first the Bayesian network corresponding to a single time step t (neglecting the index t). Let $\mathbf{q} \in \mathbb{R}^d$ refer to the configuration of the d-DOF arm of the demonstrator (during learning) or the robot (during reproduction). Let $\mathbf{x} \in \mathbb{R}^{3p+3}$ be the vector of the 3D positions of relevant objects in the scene, i.e.,

$$\mathbf{x} = \{\mathbf{x}_E, \mathbf{x}_1, \ldots, \mathbf{x}_p\}, \tag{8.1}$$

where $\mathbf{x}_1, \ldots, \mathbf{x}_p$ are the positions of the p objects in the scene. In the remainder of this chapter, we use a robot manipulator for imitating humanoid arm movements. Thus, $\mathbf{x}_E$ encodes the position of the human hand during the demonstrations and the position of the robotic end effector during the reproduction. Note that any set of body parts can be used in case full body actions should be imitated without changing the math except for adding additional variables and indices ($\mathbf{x}_E$ would become $\mathbf{x}_{E_1}, \ldots, \mathbf{x}_{E_{p'}}$ for p' body parts). Further, we denote a noisy observation of the true pose $\mathbf{x}$ as $\mathbf{y} \in \mathbb{R}^{3p+3}$, and a noisy observation of the true joint configuration $\mathbf{q}$ as $\tilde{\mathbf{y}} \in \mathbb{R}^d$.

Our DBN depicted in Figure 8.2 gives rise to the following factorization:

$$p(\tilde{\mathbf{r}}, \mathbf{q}, \tilde{\mathbf{y}}, \mathbf{r}, \mathbf{x}, \mathbf{y}) = p(\tilde{\mathbf{r}})p(\mathbf{r})p(\mathbf{q} \mid \tilde{\mathbf{r}})p(\mathbf{x} \mid \mathbf{q}, \mathbf{r})p(\tilde{\mathbf{y}} \mid \mathbf{q})p(\mathbf{y} \mid \mathbf{x}). \tag{8.2}$$

In our model, we assume the following distributions in the nodes of the DBN: the observation models $p(\tilde{\mathbf{y}} \mid \mathbf{q})$ and $p(\mathbf{y} \mid \mathbf{x})$ are assumed to be Gaussian distributions with mean $\mathbf{q}$ and $\mathbf{x}$ and a (known) variance that corresponds to the noise in the observations. Similarly, we model the distributions over constraints $p(\tilde{\mathbf{r}})$ and $p(\mathbf{r})$ as Gaussian distributions. For each time step t, we estimate the mean and covariance of these distributions from the human demonstrations. Since we have no information about the distributions over constraints in the beginning, we set their initial variance to infinity.

The posterior about the objects in the scene can be split into the forward model of the manipulator $p(\mathbf{x}_E \mid \mathbf{q})$, and the constraints between the objects in the scene and the end effector. By applying the product rule and by assuming that the poses of objects are independent from the joint configuration given the position of the end effector, we obtain

$$p(\mathbf{x} \mid \mathbf{q}, \mathbf{r}) = p(\mathbf{x}_E \mid \mathbf{q}) p(\mathbf{x}_1, \ldots \mathbf{x}_p \mid \mathbf{x}_E, \mathbf{r}). \tag{8.3}$$

By further applying the product rule, we obtain:

$$\begin{aligned} p(\mathbf{x} \mid \mathbf{q}, \mathbf{r}) &= p(\mathbf{x}_E \mid \mathbf{q}) p(\mathbf{x}_1 \mid \mathbf{x}_E, \mathbf{r}) p(\mathbf{x}_2, \ldots \mathbf{x}_n \mid \mathbf{x}_E, \mathbf{r}) && (8.4)\\ &= p(\mathbf{x}_E \mid \mathbf{q}) \prod_{i=1,\ldots,p} p(\mathbf{x}_i \mid \mathbf{x}_E, r_i) && (8.5)\\ &= p(\mathbf{x}_E \mid \mathbf{q}) \prod_{i=1,\ldots,p} p(\mathbf{x}_i - \mathbf{x}_E \mid r_i) && (8.6)\\ &\approx p(\mathbf{x}_E \mid \mathbf{q}) \prod_{i=1,\ldots,p} \mathcal{N}_{r_i}(\boldsymbol{\mu}_i; \Sigma_i) && (8.7) \end{aligned}$$

Eq. (8.4) is obtained by assuming that given the constraints $\mathbf{r}$ as well as $\mathbf{x}_E$, the positions of two objects in the scene are independent. By applying this assumption p times, we obtain Eq. (8.5). We additionally assume that the posterior about the pose of the end effector $p(\mathbf{x}_E \mid \mathbf{q})$ is Gaussian and thus corresponds to the kinematic model.

Finally, we make the assumption that the relative position of object parts to end effector position can be described using Gaussian distributions. Thus, the individual per-object constraints $r_i \in \mathbf{r}$ are represented by Gaussians with mean $\boldsymbol{\mu}_i \in \mathbb{R}^3$ and covariance $\Sigma_i \in \mathbb{R}^{3\times3}$. This leads to Eq. (8.7). Similarly, the individual joint constraints $\tilde{r}_j \in \tilde{\mathbf{r}}$ are represented by individual Gaussians with mean $\mu_j \in \mathbb{R}$ and variance $\sigma_j^2 \in \mathbb{R}$. Thus, $p(\mathbf{q} \mid \tilde{\mathbf{r}})$ can be computed as

$$\begin{aligned} p(\mathbf{q} \mid \tilde{\mathbf{r}}) &= \prod_{j=1,\ldots,d} p(q_j \mid \tilde{r}_j), && (8.8)\\ &= \prod_{j=1,\ldots,d} \mathcal{N}_{\tilde{r}_i}(\tilde{\mu}_i; \tilde{\sigma}_i^2) && (8.9) \end{aligned}$$

With this, we have a full specification of the DBN depicted in Figure 8.2. We learn Gaussian distributions in Eq. (8.7) and Eq. (8.9) for the joint constraints

$\tilde{\mathbf{r}}$ and Cartesian constraints $\mathbf{r}$ gathered from multiple human demonstrations. Further, we assume that the kinematic model $p(\mathbf{x} \mid \mathbf{q})$ of the robot is known, as well as the observation models $p(\tilde{\mathbf{y}} \mid \mathbf{q})$ and $p(\mathbf{y} \mid \mathbf{x})$.

So far, we have only considered the Bayesian network corresponding to a single time step t. We extend this to multiple time steps by copying the template DBN in Figure 8.2 for each time step. Furthermore, we add edges between consecutive poses $\mathbf{x}^t$ and $\mathbf{x}^{t+1}$ and joint configurations $\mathbf{x}^t$ and $\mathbf{x}^{t+1}$ to ensure temporal consistency and to track the poses and configurations from human demonstrations with a Kalman filter.

8.1.1 Learning Task Descriptions from Human Demonstrations

During learning, the robot observes a person that repeatedly carries out the task the robot has to perform. Given the DBN structure explained above, the key challenge of this learning phase is to learn the constraints between objects and the manipulator $p(\mathbf{r})$ and, if needed, the constraints on joint configurations $p(\tilde{\mathbf{r}})$.

Motion Capturing and Object Pose Estimation

To estimate the motion trajectories of the human demonstrator while executing a task and the 3D position of relevant objects in the scene, we use passive markers and images of a monocular camera (Fiala, 2005). We attach compounds of four markers around the teacher's arm (see Figure 8.1a) to bypass the problem of occlusions. In most cases, not more than one marker of the same compound is visible. To deal with the case that two markers are visible simultaneously (sensing more than two markers at the same time is impossible due to their mutual orthogonality within one compound), we perform a linear interpolation between their poses depending on the degree of visibility of the markers. Finally, we apply a Kalman filter to track the 3D marker position estimates over time. To derive the demonstrator's joint angles from marker poses, we use an anthropomorphic arm model and apply straightforward geometric operations. As a result, we are able to reliably estimate the arm configuration $\mathbf{q}^{t,g}$ and object poses $\mathbf{x}^{t,g}$ for each time step $t = 1, \ldots, T$ in demonstration $g = 1, \ldots, G$, where G is the number of demonstrations.

Aligning Multiple Demonstrations

Our approach relies on multiple demonstrations to achieve a good generalization. One problem when generalizing task descriptions from multiple demonstrations is the fact that the observations of the individual demonstrations are not time-synchronized, even though the different demonstrations typically do not vary largely. To deal with varying movement velocity profiles, we apply derivative dynamic time warping (Keogh and Pazzani, 2001) which is able to account for local distortions in the time domain by computing a nonlinear transformation of the time axis of the individual demonstrations. Based on the aligned demonstrations, we can derive the constraints $p(\mathbf{r})$ and $p(\tilde{\mathbf{r}})$ which encode the action to imitate.

Learning the Task Constraints

By assuming that the constraints $p(\mathbf{r})$ and $p(\tilde{\mathbf{r}})$ are composed of individual Gaussians, we can directly infer a mean and a variance estimate of the individual constraints for each point in time given the estimates for $p(\mathbf{x})$ and $p(\mathbf{q})$. The estimated variance in each of the constraints is of particular importance since it describes how accurately the demonstrator enforced a particular constraint during his demonstrations.

Formally, an object-manipulator constraint r_i is fully described by the mean position of the objects relative to the manipulator $\boldsymbol{\mu}_i \in \mathbb{R}^3$ and a covariance matrix $\Sigma_i \in \mathbb{R}^{3\times3}$.

In theory, we could compute $\boldsymbol{\mu}_i$ and Σ_i directly from the estimates for $\mathbf{x}$ and $\mathbf{q}$ during learning. In practice, however, we typically have to deal with a few demonstrations and, therefore, rather rough and non-smooth estimates are obtained if the values are computed directly. To overcome this problem, we apply a Parzen window kernel estimator for computing smooth function approximations. This is a nonparametric technique that allows to estimate a value for $\boldsymbol{\mu}$ based on a set of sample points. We weight each training sample by a factor

$$w_i^{t,g}(t') = \frac{1}{\eta^t} K\left(\frac{t'-t}{h}\right), \tag{8.10}$$

where h is the Parzen window size (empirically determined, $h = 0.2\,\mathrm{s}$) and K is a kernel function. We use the standard choice for the K, namely the squared exponential kernel

$$K(d) = \exp\left(-\frac{1}{2}\left\|d\right\|^2\right). \tag{8.11}$$

Here, η^t is a normalizing constant, to ensure that all sample weights with respect to time step t sum to one, i.e.,

$$\eta^t = \sum_{g=1}^{G}\sum_{t'=1}^{T} w^{t,g}(t'), \tag{8.12}$$

where G stands for the number of demonstrations. This gives us for the constraint r_i the weighted sample mean

$$\boldsymbol{\mu}_i^t = \sum_{g=1}^{G}\sum_{t'=1}^{T} w^{t,g}(t')\left[\mathbf{x}_i^{t,g} - \mathbf{x}_E^{t,g}\right], \tag{8.13}$$

between object i in the scene and the end effector of the manipulator. Similarly, the weighted sample covariance can be estimated as

$$\Sigma_i^t = \frac{\sum_{g=1}^{G} \sum_{t'=1}^{T} w_i^{t,g}(t') \left(\left[\mathbf{x}_i^{t,g} - \mathbf{x}_E^{t,g} \right] - \boldsymbol{\mu}_i^t \right) \left(\left[\mathbf{x}_i^{t,g} - \mathbf{x}_E^{t,g} \right] - \boldsymbol{\mu}_i^t \right)^T}{1 - \sum_{g=1}^{G} \sum_{t'=1}^{T} \left(w_i^{t,g}(t') \right)^2}. \tag{8.14}$$

This procedure is carried out for each time step t and each object i in the scene. Accordingly, we specify a configuration constraint as a normal distribution describing a mean configuration $\tilde{\boldsymbol{\mu}}^t \in \mathbb{R}^d$ and an associated covariance matrix $\tilde{\Sigma}^t \in \mathbb{R}^{d \times d}$. Similar to the object-manipulator constraints, we compute the joint constraints $\tilde{\mathbf{r}}^t$ as

$$\tilde{\boldsymbol{\mu}}^t = \sum_{g=1}^{G} \sum_{t'=1}^{T} w_i^{t,g}(t') \mathbf{q}^{t,g} \tag{8.15}$$

and

$$\tilde{\Sigma}^t = \frac{\sum_{g=1}^{G} \sum_{t'=1}^{T} w_i^{t,g}(t') \left(\mathbf{q}^{t,g} - \tilde{\boldsymbol{\mu}}^t \right) \left(\mathbf{q}^{t,g} - \tilde{\boldsymbol{\mu}}^t \right)^T}{1 - \sum_{g=1}^{G} \sum_{t'=1}^{T} \left(w_i^{t,g}(t') \right)^2}. \tag{8.16}$$

8.1.2 Reproducing Tasks

The goal of the reproduction or imitation phase is to carry out the demonstrated task to achieve the same result. Given the DBN, we can seek for the configuration of joints $\mathbf{q}^*$ that maximizes the likelihood given the learned task model.

Incremental Optimization

If we consider only a single time step of the reproduction, we seek for the configuration $\mathbf{q}^*$ that maximizes the joint probability of a single time slice. After the learning phase, the task space constraints $p(\tilde{\mathbf{r}})$ and the configuration space constraints $p(\mathbf{r})$ are known. Furthermore, the robot can control its manipulator by specifying a joint configuration $\mathbf{q}$ and does not have to rely on noisy marker observations $\tilde{\mathbf{y}}$ as in the learning phase. Therefore, the maximization turns into

$$\mathbf{q}^* = \arg\max_{\mathbf{q}} \; p(\mathbf{q} \mid \tilde{\mathbf{r}}) p(\mathbf{x} \mid \mathbf{q}, \mathbf{r}). \tag{8.17}$$

As discussed in Section 8.1, the posteriors $p(\mathbf{q} \mid \tilde{\mathbf{r}})$ and $p(\mathbf{x} \mid \mathbf{q}, \mathbf{r})$ are basically products of Gaussians and lead to a Gaussian distribution again. Thus, to maximize the joint probability, we need to determine the mean of this Gaussian distribution. To do so, we proceed as follows. Consider that we are currently at time step t and seek for the joint configuration that maximizes Eq. (8.17) at $t+1$. Each constraint between an object i and the end effector generates a relative displacement vector $\boldsymbol{\Delta}_i$, i.e,

$$\boldsymbol{\Delta}_i^{t+1} = \boldsymbol{\mu}_i^{t+1} - (\mathbf{x}_i^t - \mathbf{x}_E^t). \tag{8.18}$$

Correspondingly, each object-manipulator constraint has an associated covariance matrix Σ_i^{t+1} that encodes how much variation in this constraint was observed in the human demonstrations. One can imagine the underlying idea here visually as follows: the end effector is pulled towards each constraint according to the variance present in the demonstrations. Constraints with low variance will influence the motion of the manipulator stronger than constraints with high variance.

Since we want to compute a new joint configuration for the robot, we need to convert the constraints expressed in world coordinates in joint space. We achieve this by applying a variant of the damped-least squares method described by Buss and Kim (2005). This approximative technique performs a linearization of the kinematic function. According to this method, a desired movement in Cartesian space ($\boldsymbol{\Delta}$) is transformed to an executable movement in joint space ($\tilde{\boldsymbol{\Delta}}$) by

$$\tilde{\boldsymbol{\Delta}}_{\mathbf{r}_i}^{t+1} = J\left(JJ^T + \lambda^2 I\right)^{-1} \boldsymbol{\Delta}_i^{t+1}, \tag{8.19}$$

$$\tilde{\Sigma}_{\mathbf{r}_i}^{t+1} = \left(J\left(JJ^T + \lambda^2 I\right)^{-1}\right) \Sigma_i^{t+1} \left(J\left(JJ^T + \lambda^2 I\right)^{-1}\right)^T, \tag{8.20}$$

where λ is the so-called damping factor and J refers to the Jacobian of the end effector. Due to the linear mapping, we also obtain a Gaussian in configuration space. The configuration space constraints defined by $\tilde{\mathbf{r}}^{t+1}$ can easily be used to compute an executable movement

$$\tilde{\boldsymbol{\Delta}}_{\tilde{\mathbf{r}}}^{t+1} = \tilde{\boldsymbol{\mu}}^{t+1} - \mathbf{q}^t, \tag{8.21}$$

$$\tilde{\Sigma}_{\tilde{\mathbf{r}}}^{t+1} = \tilde{\Sigma}^{t+1}. \tag{8.22}$$

Note that the covariance matrix of the configuration constraints does not need to be transformed as it is already expressed in configuration space.

All constraints resulting from the observation of the demonstrator's configurations or from the arrangement of objects in the scene are now expressed in terms of updates in configuration space. This allows us to merge these normal distributions into a single Gaussian (Calinon and Billard, 2008). The resulting distribution is the product over the $p+1$ Gaussians resulting from the p task space relations plus the joint space relations. We can use it to directly obtain an estimate $\hat{\mathbf{q}}^{t+1}$ of the next configuration $\mathbf{q}^{*,t+1}$ according to Eq. (8.17) by selecting the mean from this combined Gaussian, as given by

$$\hat{\mathbf{q}}^{t+1} = \mathbf{q}^t + \hat{\Sigma}^{t+1}\left((\tilde{\Sigma}_{\tilde{\mathbf{r}}}^{t+1})^{-1}\tilde{\Delta}_{\tilde{\mathbf{r}}}^{t+1} + \sum_{i=1}^{n}(\tilde{\Sigma}_{\mathbf{r}_i}^{t+1})^{-1}\tilde{\Delta}_{\mathbf{r}_i}^{t+1}\right), \tag{8.23}$$

with

$$\hat{\Sigma}^{t+1} = \left(\hat{\Sigma}^t + (\tilde{\Sigma}_{\tilde{\mathbf{r}}}^{t+1})^{-1} + \sum_{i=1}^{n}(\tilde{\Sigma}_{\mathbf{r}_i}^{t+1})^{-1}\right)^{-1}. \tag{8.24}$$

The mean of this distribution specifies the configuration of the robot at the next time step that maximizes the probability distribution given in Eq. (8.17).

Local Optimization with Obstacles

The technique described in the previous section can directly be applied to deal with unforeseen obstacles in the scene during reproduction. Consider that the robot observes an obstacle during the imitation that was not there during the demonstrations. To avoid this obstacle during the reproduction task, we can add additional constraints between the observed obstacle and the closest point on the robot's body as used in approaches based on potential fields for collision avoidance.

Without changing the framework described above, the robot can reactively introduce constraints for avoiding obstacles while carrying out its task as similarly as possible to the human demonstrations. Let $\mathbf{x}_O$ be the position of the obstacle. Instead of adding a repellent force, we add an attractor at the opposite side of the end effector,

$$\mathbf{\Delta}_O = -\alpha\frac{\mathbf{x}_O - \mathbf{x}_E}{\|\mathbf{x}_O - \mathbf{x}_E\|} \tag{8.25}$$

$$\Sigma_O = \beta \cdot \exp\left(\|\mathbf{x}_O - \mathbf{x}_E\|^2\right) \cdot I, \tag{8.26}$$

where α determines the desired distance to the obstacle and β gives the relative importance with respect to the other constraints.

It should be noted that this technique works well for small or rather simple structured obstacles added to the scene. In case complex or, for example, U-shaped obstacles are found in the environment, this approach is likely to suffer from local minima caused by contradictory constraints.

Global Optimization

The problem of local minima, however, can be avoided by globally optimizing the joint probability distribution of the DBN over all time steps $1 \ldots T$ of the task sequence at once, i.e.,

$$\mathbf{q}^{*,1:T} = \arg\max_{\mathbf{q}^{1:T}} p(\tilde{\mathbf{r}}^{1:T}, \mathbf{q}^{1:T}, \tilde{\mathbf{y}}^{1:T}, \mathbf{r}^{1:T}, \mathbf{x}^{1:T}, \mathbf{y}^{1:T}). \tag{8.27}$$

Note that at a particular time step t, only the first $1 \dots t$ observations $\mathbf{y}^{1:t}$ are already available and can be included for planning. Doing this optimization on a global level, however, comes with significantly increased computational cost due to the high dimensionality of $\mathbf{q}^{*,1:T}$.

One way of efficiently estimating $\mathbf{q}^{*,1:T}$ is to make use of probabilistic roadmaps or rapidly-exploring random trees (LaValle, 2006) to find the shortest path using A^* on the sampled set of nodes. Given that we properly encode the likelihoods of all constraints in the cost function later used by A^*, the solution of the planner will approximate Eq. (8.27) well.

As cost function, we use the Mahalanobis distance of the combined Gaussian distribution $\mathcal{N}(\hat{\mathbf{q}}^t; \hat{\Sigma}^{t+1})$ as computed in Eq. (8.23) and Eq. (8.24). This distribution incorporates all constraints $\mathbf{r}$, $\tilde{\mathbf{r}}$ and the obstacle constraints in a time-dependent way. For a configuration $\mathbf{q}$ at time $t+1$, we define the cost as the likelihood with respect to the previously computed combined distribution, i.e.,

$$cost(\mathbf{q}^{t+1}) = \left(\mathbf{q}^t - \hat{\mathbf{q}}^{t+1}\right)^T (\hat{\Sigma}^{t+1})^{-1} \left(\mathbf{q}^t - \hat{\mathbf{q}}^{t+1}\right). \tag{8.28}$$

Then, finding the cost-optimal sequence of configurations $\mathbf{q}^{*,1:T}$ is equivalent to maximizing the likelihood of the trajectory $\mathbf{q}^{*,1:T}$ in Eq. (8.27).

8.2 Experiments

We carried out a set of experiments to evaluate our approach. We observed a human demonstrator equipped with markers of the ARToolkit as depicted in Figure 8.3 at a rate of 5 Hz. We used this data to learn the constraints for the task reproduction. In our experiments, we segmented the training trajectories corresponding to the individual demonstrations manually. Further, instead of estimating the full covariance matrices as described in Section 8.1.1, we restricted ourselves to the diagonal values. We found that this makes the estimation of the task constraints more robust when the number of expert demonstrations is limited.

The goal of our experiments is to demonstrate that different manipulation tasks can be learned with our approach. We show this using the example of different pick-and-place tasks and the task of cleaning a white board. In these experiments, the robot successfully generalized the learned task models to novel spatial setups. Further, we evaluate the convergence behavior of our approach with respect to the number of required human demonstrations. Lastly, we show that global planning can be used to circumvent local minima during task reproduction, for example, in the presence of contradictory constraints.

8.2.1 Imitating Human Actions

To imitate the observed behavior, we reproduced the tasks using a real robot equipped with a manipulator and two simulated robots, one with a manipulator and one with a human-like arm.

In our first experiment, a human demonstrator repeated a simple pick-and-place task four times, as illustrated in Figure 8.3. In this experiment, we tracked the configuration of the human arm, the Cartesian position of the human hand, as well as the positions of the mug and the table. Figure 8.4 shows the reproduction of the pick-and-place task after being demonstrated four times. The human-like simulated robot considers both the joint and the task constraints, which leads to the fluent, human-like movement.

In Figure 8.5, the same task was reproduced by the robotic manipulator in simulation. Since the demonstrator and the imitator have significantly different kinematics, we disabled the joint constraints $\tilde{r}$ during reproduction. Furthermore, in this experiment, we swapped the positions of the two tables. As can be seen from the plotted trajectories, the robot was able to generalize the task successfully, i.e., it approached the tables in the correct order.

In the experiment depicted in Figure 8.6, we analyzed the number of demonstrations needed until a task could reliably be reproduced. For this analysis, a teacher picked up a cup and placed it at a distance of 1 m. After the first demonstration, the robot could not sufficiently generalize the task to reproduce it reliably. After the second demonstration, the robot correctly approached the picking region but slightly missed the spot for placing the cup. After four demonstrations, the task model converged and the robot was able to reproduce the task reliably.

8.2.2 Dealing with Obstacles during Imitation

The additional obstacle constraints described in Section 8.1.2 allow the robot to deal with unforeseen obstacles during task execution. The obstacle constraints act similar to a potential field pushing the robot away from obstacles. We implemented the perception of obstacles using additional ARToolkit markers.

Figure 8.7 illustrates an example for a constraint in a pick and place task. The figure shows the reproduced trajectory for the obstacle free case and a trajectory that was generated in the presence of an obstacle. As can be seen, the robot moved its arm over the obstacle in order to avoid a collision.

We carried out a white board cleaning task that nicely illustrates the properties of the presented methods. First, a human repeatedly cleaned a white board in an area bounded by 4 markers with the same number of ups and downs (see Figure 8.1a). Then, we attached a sponge to the robot and let it imitate the demonstrated task. In the first experiment, we modified the size of the area to clean for illustrating the capabilities of generalization. Photos from this experiment can be seen in Figure 8.8. Note that in case the area to clean is much larger than during learning, the white board may not

Fig. 8.3 Human demonstration of a pick-and-place task. The trajectory is recorded both in task and joint space.

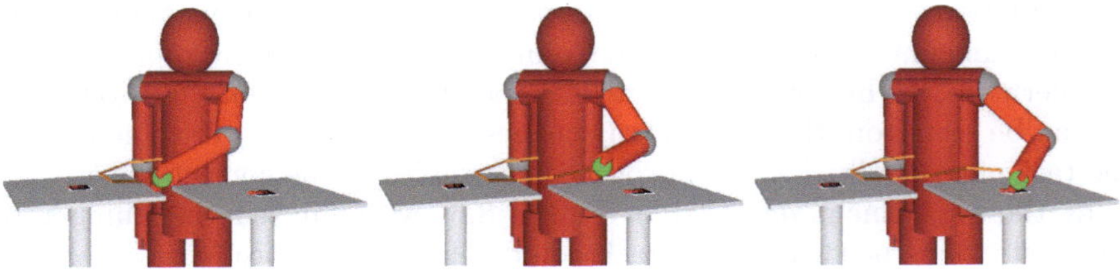

Fig. 8.4 Reproduction of the pick-and-place task by a human-like manipulator in simulation using both task and joint space constraints.

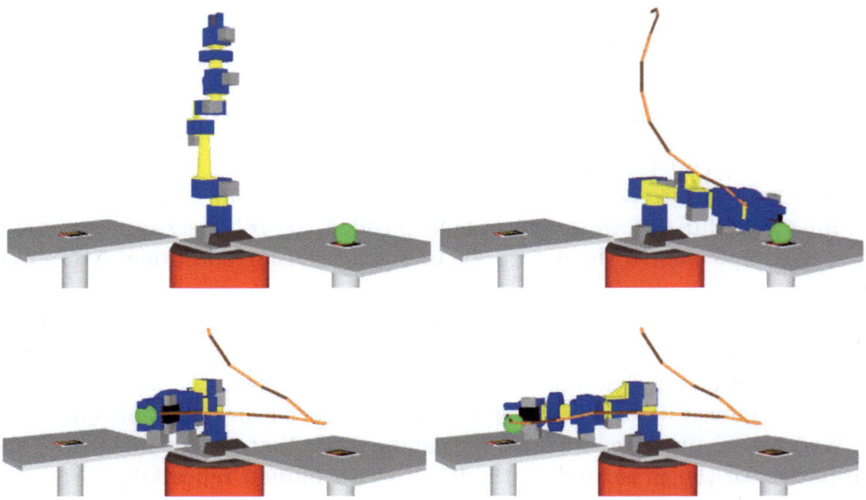

Fig. 8.5 Reproduction of the same pick-and-place task by our 6-DOF manipulator. Note that the robot successfully generalized the task as the start and goal location were exchanged.

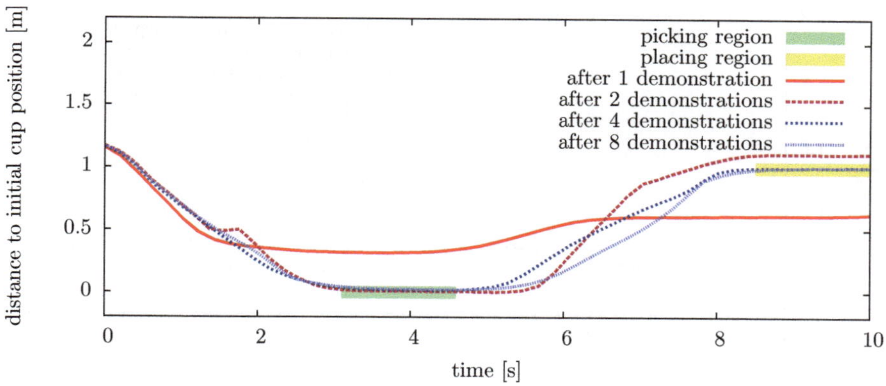

Fig. 8.6 Evaluation of the convergence behavior of our approach with respect to the number of expert demonstrations.

be cleaned well. The reason for that is that our approach imitates the task at a trajectory level and thus does not generalize coverage patterns over areas. As a result, there might be parts of the white board that are not covered by the imitated trajectory, and thus, will not be cleaned.

In the second experiment, we introduced an additional obstacle marker during reproduction (see Figure 8.9(a)). As a result, the robot imitated the cleaning task while avoiding obstacles (and thus not cleaning the area of the marker). For reasons of illustration, we removed the marker during the experiment but kept it in the internal memory of the robot. As a result, the robot did not clean the corresponding area. Four photos were taken during the reproduction and are depicted in Figure 8.9. In our experiment, the robot lifted the sponge away from the white board (in the direction of the observing camera) in order to avoid the area.

It should be noted that the degree of generalization of our approach strongly depends on amount of the variation present in the expert's demonstrations. If only few variations exist, the learned task model might over-fit to these demonstrations. On the other hand, too much variation makes it harder for the robot to find the invariants. A good solution here is to teach a novel manipulation skill to a robot incrementally, i.e., demonstrating it a few times and checking how well the robot has generalized the task.

8.2.3 Imitation by Planning

In the experiments presented above, we applied the incremental strategy to reproduce the task. This can be solved efficiently online, but this strategy suffers from local minima, for example, in the presence of U-shaped obstacles. Such an example is presented in Figure 8.10 where the robot gets stuck when using the incremental strategy.

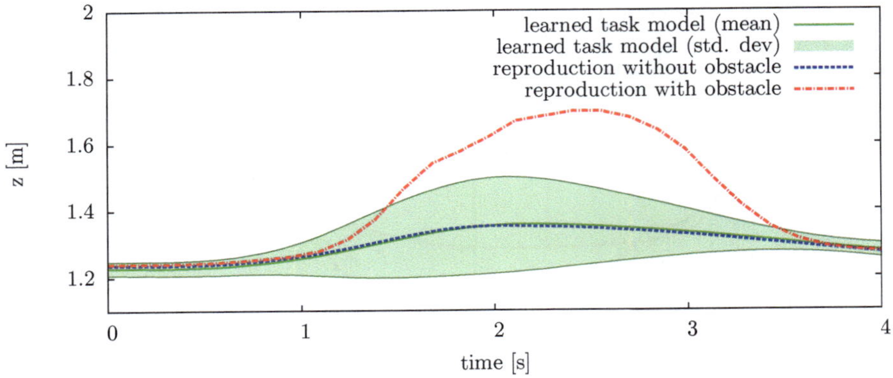

Fig. 8.7 Visualization of the learned constraints including the variances as well as two reproduction trajectories – one for the obstacle free case and one in case an obstacle blocks the trajectory.

If one applies the global optimization technique described in Section 8.1.2, one can overcome this problem since the optimal solution over all time steps is computed. Thus, the robot is able to reproduce the task including the avoidance of the U-shaped obstacle. This global method, however, comes with a significantly increased computational load.

To summarize our results, we demonstrated in our experiments that the learned task descriptions are general enough so that the robot can successfully reproduce the task even when the positions of the objects in the scene were changed. Furthermore, we showed that a robot can include additional constraints during task reproduction, for example, to evade obstacles that were not there during learning. In convergence experiments, we found that our approach required only four demonstrations to reliably reproduce a pick-and-place task. Finally, we showed that global planning finds solutions where incremental approaches gets stuck.

8.3 Related Work and Discussion

Various techniques have been proposed in the past to transfer a task description to robotic systems. In the industrial context, a common solution is to teach suitable trajectories using a joystick or kinesthetic training. Assuming that the environment is precisely specified, the robot can exactly reproduce the recorded sequence and no generalization is required. However, if the observations are noisy or unpredicted disturbances in the task environment occur, simple playback of the recorded motion is not sufficient to reliably reproduce a given task. Detailed surveys on the current state-of-the-art in

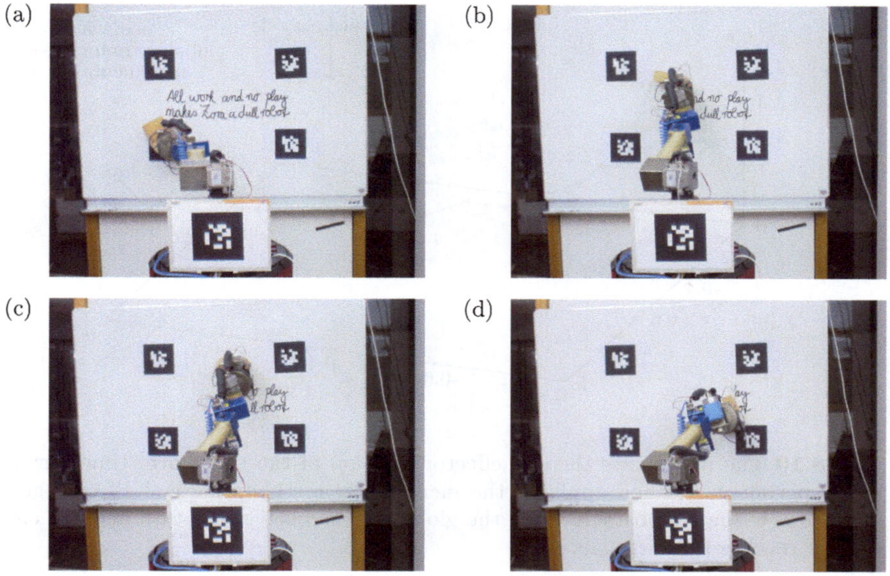

Fig. 8.8 The reproduction of the board cleaning task by our robot. It imitates the zig-zag movement for cleaning the board with the sponge.

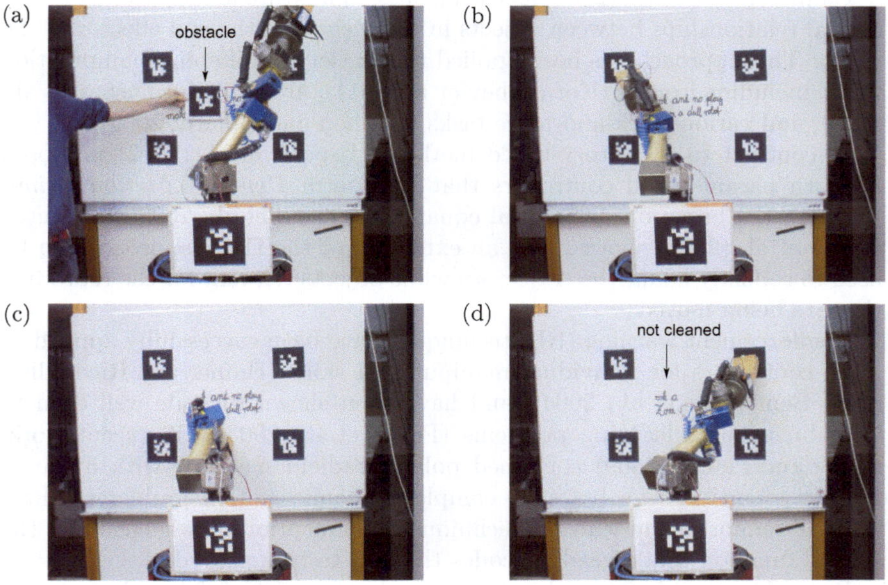

Fig. 8.9 During the second reproduction, we introduced an additional obstacle that the robot has to consider using a visual marker. As a result, the robot does not clean the area under the obstacle.

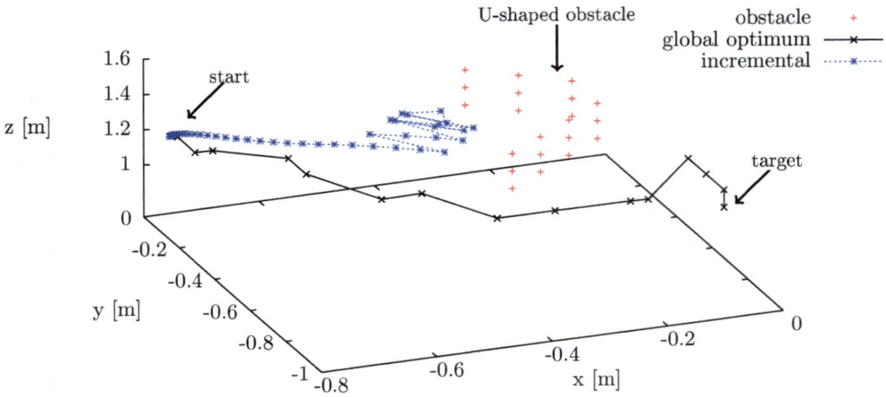

Fig. 8.10 The plot shows the end effector position of the robot over time during two experiments. When applying the incremental method, the end effector gets stuck in a U-shaped obstacle while the global method solves the task and the end effector reaches the target location.

the domain of imitation learning and robot programming by demonstration are given in the works of Billard et al. (2008) and Argall et al. (2009).

Calinon and Billard (2008) learned Gaussian mixture models to encode the spatial relationships between objects in the scene and the end effector of the robot. This approach has been applied to a variety of different manipulation tasks, including ironing (Kormushev et al., 2011), archery (Kormushev et al., 2010), and various pick-and-place tasks (Calinon and Billard, 2009).

In contrast to trajectory-based methods, Ijspeert et al. (2002) proposed to learn parametrized controllers that they term *Dynamic Motion Primitives* (DMP) based on differential equations (Schaal et al., 2003). Recently, Pastor et al. (2009) showed how an extension of the DMP approach can be used to robustly adapt the trajectory while objects relevant for the reproduction are being moved.

Reinforcement learning (RL) techniques have been successfully applied to learn controllers for individual manipulation skills (Hafner and Riedmiller, 2007; Bentivegna et al., 2004), and have been shown to scale well even to high-dimensional learning problems (Peters et al., 2003). In recent work, Kober and Peters (2009) combined policy gradient methods with dynamic motion primitives to solve more complex dynamic motion problems. However, all reinforcement learning techniques assume prior knowledge about the reward function that already encodes the goal to be achieved.

Recently, several approaches have been proposed to infer such a reward function from demonstrations of an expert using *inverse reinforcement learning* (Abbeel et al., 2007; Ratliff et al., 2007; Ziebart et al., 2008). Depending on the underlying model, the controller can be learned from imperfect expert demonstrations and the resulting policy can even outperform the expert (Coates et al., 2008).

Other authors proposed to use hidden Markov models (HMM) for encoding the temporal sequence of manipulation goals (Asfour et al., 2006; Calinon et al., 2005; Tso and Liu, 1996). Pardowitz and Dillmann (2007) presented a system that generalizes over household tasks in a hierarchical manner. Actions performed by the human demonstrator are recognized as a sequence of "elementary operators", of which a graph-based task representation is learned. In this approach, the incrementally updated network topology reflects the learned temporal ordering of the individual actions. Although not directly related to imitation learning, Beetz et al. (2010) described a system that enabled a robot to retrieve semantic task instructions from the world wide web in order to efficiently set a table.

While symbolic representations are well suited for planning and reasoning, their limitation to higher-level skills renders them inapplicable in domains where a continuous motor control is required. By contrast, trajectory learning directly starts by encoding each demonstration by a sequence of continuous observations. Various approaches exist for iterative trajectory optimization (Ratliff et al., 2009). Due to the high-dimensional input space, dimensionality reduction techniques are often applied. Chalodhorn et al. (2007) used principal component analysis (PCA) to reduce the high-dimensional motion capture data of a recorded human walk. While a direct playback of the human data on a humanoid robot would make it fall, the authors showed that after a few trials the robot was able to modify the imitated gait incrementally. Similarly, Grimes et al. (2006) also used PCA to reduce the high-dimensional configuration space and applied a DBN to infer dynamically stable imitative actions using constraint variables and a learned forward model of the robot dynamics. Grochow et al. (2004) consider the pose ambiguity problem that arises in inverse kinematics as a constrained optimization task. They show that different movement styles can be learned from human demonstrations and used as a prior of an animated character.

The approach presented in this chapter is inspired by the prior work of Calinon and Billard (2008). In contrast to their approach, we (1) provide a probabilistic formulation of the imitation learning problem as a dynamic Bayesian network, (2) show that this formulation facilitates the inclusion of additional constraints, and (3) demonstrate that global optimization during task reproduction provides solutions where incremental approaches fail. It should be noted that our method does not generalize the learned tasks above the trajectory level. For example with the white-board cleaning task, our method always reproduces the same number of ups and downs, which means that – in its current form – cannot deviate to generate other coverage plans. Such high-level generalization capabilities are clearly also highly relevant in many real-world applications. One possible solution is to apply imitation learning at different levels simultaneously, i.e., to combine low-level motor learning, trajectory-level imitation learning for atomic actions, and high-level task learning. This, however, remains a topic for future investigation.

8.4 Summary

In this chapter, we developed an approach to imitation learning that enables a robot to learn, generalize, and reproduce tasks by observing a human demonstrator. We model the description of a manipulation task as a DBN in which special nodes encode the geometrical relationships between objects in the scene and the end effector of the robot. To reproduce the task, we seek for the action sequence that maximizes the likelihood of the DBN. The formulation as a DBN allows to flexibly add or remove constraints during task reproduction, for example, to avoid obstacles. In experiments carried out in simulation and on a real robot, we demonstrated that a robot using our approach can learn, generalize, and reproduce various manipulation tasks even under different spatial setups. To conclude, our approach provides a solution that enables normal users to intuitively teach novel tasks to a service robot.

Chapter 9
Conclusions

In this book, we presented several innovative techniques that enable mobile manipulation robots to robustly operate in unstructured environments under changing, real-world conditions, which is essential for the success of mobile manipulation robots in the future. Many of the relevant applications require that robots function robustly in new situations while they are dealing with considerable amounts of noise and uncertainty. Therefore, the main objective of this work was to develop novel approaches that enable manipulation robots to autonomously acquire the models they need to successfully implement their service tasks.

In domestic environments, a manipulation robot needs to operate over extended periods of time without being supervised by an expert. Therefore, we investigated methods which a robot can use to learn its body schema from scratch and adapt it in case of changes. We introduced a flexible representation for kinematic models based on Gaussian processes and Bayesian networks and devised an efficient algorithm that recovers both the kinematic structure and the kinematic properties of the manipulation robot. Our approach enables the robot to position its end effector accurately even in the presence of hardware failures. In our experiments, we demonstrated that self-observation allows a robot to significantly increase its mean time between failures.

A central task of service robots is to interact with articulated objects, for example, to open doors in order to navigate between rooms or to pick up objects from cabinets or drawers. To allow the robot to deal with such objects, we extended our approach to kinematic model learning to become applicable also to passively-actuated articulated objects. Our proposed framework combines parametric and nonparametric models and provides a principled solution to compare and rank alternate models. Furthermore, we showed that prior knowledge can consistently be used during model learning. As we demonstrated in a large set of experiments, robots using our approach can learn accurate kinematic models for a large number of different articulated objects, and two different mobile manipulators could robustly operate various real-world objects. Furthermore, we developed a marker-less perception system to visually detect articulated objects in kitchen environments and to learn their kinematic models. With our approach, we provide a complete probabilistic framework that enables robots to learn and operate various types of articulated objects.

J. Sturm: *Approaches to Probabilistic Model Learning*, STAR 89, pp. 179–183.
DOI: 10.1007/978-3-642-37160-8_9

In addition to articulated objects, service robots also need to manipulate many other objects such as bottles, silverware, or dishes. We showed that robots can benefit from tactile sensing, for example, to verify that it has grasped the correct object and to determine its content. We developed a system based on the bag-of-features approach where the robot generates a tactile vocabulary to learn a codebook for recognizing objects. In our experiments, we demonstrated that a robot using our approach can distinguish between a large set of objects including both typical household objects and industrial work pieces. Additionally, we addressed the problem of estimating the internal state during object manipulation, which enables a service robot, for example, to determine whether a bottle contains liquid. The robot achieves this task by extracting high-frequency features from the tactile signal and by learning a decision tree classifier. Our results indicate that tactile sensing is a useful source of information for a robot to augment its perceptions during object manipulation.

Another prerequisite for successful service robotics applications is that normal users can quickly and intuitively instruct the robot to perform novel tasks. Inspired by work on imitation learning, we developed a technique to infer task descriptions from human demonstrations. As imitation learning is a high-dimensional learning problem, we approach the problem by factorizing it using dynamic Bayesian networks into individual task constraints that can be learned separately from the data. As the constraints are expressed relative to other objects in the scene, a robot can reproduce the task also in different spatial setups. In contrast to existing approaches, our solution allows to add novel task constraints dynamically during execution, for example, to avoid obstacles. The ability of our system to quickly learn novel tasks from the user is an essential feature for the everyday use of mobile manipulation robots.

All techniques presented in this book have been implemented and thoroughly tested. The experiments have been carried out in simulation as well as on real robots. We used mobile manipulation robots from Schunk, Meka, and Willow Garage. Every single approach presented in this book has been evaluated in extensive sets of real-world experiments. We demonstrated that our techniques enable robots to autonomously learn suitable models from noisy observations and use them to reliably fulfill their manipulation tasks. Our experiments support the claim that our approaches seriously decrease the dependency on hand-crafted models and significantly increase the flexibility and robustness of mobile manipulation robots.

The contributions of this work are solutions to various challenging problems in the context of model learning, imitation learning, and tactile sensing. All techniques have either been integrated into the robot operating system ROS or the robotics toolkit Carmen. Our solutions enable robots to autonomously answer the following questions:

- How can a manipulation robot position its end effector accurately, even in the presence of hardware failures?

- How can articulated objects such as doors and drawers be moved, and how can a robot operate them reliably?
- How can a manipulation robot use its tactile sensors to gain information about the objects it manipulates?
- How can a user quickly teach novel manipulation tasks to a robot, and how can the robot generalize them and reproduce them in a new situation?

We believe that the approaches presented in this book enable manipulation robots to operate in more realistic environments and we hope that the proposed solutions are relevant for future service robots that assist us in our everyday life.

9.1 Future Work

Despite the promising results presented in this book, there are several open research questions that remain for future investigation. For example, we think that the ability of a robot to observe the outcome of its actions is the key to create more dependable and resilient machines. A first step could be to apply our approach on body schema learning to other components of a mobile service robot. With a few extensions, the robot's base or an articulated sensor head could be included into our framework. Another extension is to learn additional models for specific tasks that are carried out only in a small area of the work space. For example, it could make sense to learn a specific body schema for high-precision assembly tasks on a workbench that reflects the kinematics of the robot more accurately than the global body schema describing the whole configuration space. It would also be interesting to see whether and how our approach can be generalized to learn the kinematic models of real actuators with complicated nonlinear effects, such as for example belt gear or gear backlash.

In our current approach, we learn the kinematic models from static pose observations. It would be interesting to include the velocities or accelerations of object or body parts. This would allow the robot to learn the dynamic parameters as well and enable it to plan time-optimal motion trajectories (Hollerbach et al., 2008). A dynamical model would enable the robot to accurately execute motions at higher speeds. Furthermore, a robot that can measure forces and torques while actuating an object could additionally learn friction and damping profiles and include this information in the learned model as well. A starting point for this integration could be the work of Jain et al. (2010), who studied force profiles of hundreds of doors and drawers in private residences. The robot could evaluate these profiles to estimate the torque required to open a drawer, prevent to apply excessive forces, and detect, whether the drawer is jammed. Another open question is whether knowledge about the physical structure of the world can be exploited during model learning. For example, prior knowledge about common types of mechanical linkages could help a robot to infer and disambiguate the kinematic structure. For example, the robot could exploit that drawers are more likely to be connected to the cabinet than to each other, or that a door is more likely to be attached to the door frame and not to the floor. Both cases are currently not distinguishable with our current approach.

As we have demonstrated in this book, tactile sensing provides valuable information about the state of an object. We are convinced that tactile information can support object manipulation in a variety of ways. In particular, approaches that tightly combine perception with control bear a large potential. Especially during grasping, tactile sensors provide many details about the contact state of hand and object, or object and table, that are difficult to access using other sensor modalities. For example, a robot could choose the approach trajectory that is expected to minimize the pose uncertainty while grasping the object and filtering its pose using tactile sensing. Another example is to use tactile sensors to detect that an object has made firm contact with a surface before releasing it.

Our approach on imitation learning enables a person to quickly teach novel manipulation tasks to a manipulation robot by demonstration. An interesting extension is interactive teaching: the robot tries to reproduce a task directly after the first demonstration, while the instructor incrementally provides additional demonstrations when the reproduction fails or is unsatisfactory. This would also provide feedback to the user how well the robot has inferred and generalized the task description. During the autonomous reproduction of a task, it would be beneficial if the robot was able to detect failures autonomously. This is not trivial, as, at the same time, the robot should be able to generalize a task to different situations which explicitly requires some derivations from the original demonstrations. A solution would be that the robot learns a classifier based on additional user demonstrations to decide which deviations are acceptable and which are not. Finally, a general shortcoming of trajectory-based imitation learning methods is that they do not generalize the task very well above the trajectory level. Therefore, it would also be interesting to investigate methods that focus more on the learning of higher-level task descriptions.

Several ongoing research projects are currently using or extending our approach on learning kinematic models of articulated objects. The RoboEarth project[1] aims at the creation of a worldwide object database and plans to annotate articulated objects with the models learned using our approach. The goal of the SFB/TR 8[2] is to investigate the cognitive foundations for human-centered spatial assistance systems, and plans in project A8 to extend our approach to learn 3D models of the rigid parts of articulated objects. The

[1] RoboEarth is part of the Cognitive Systems and Robotics Initiative from the European Union Seventh Framework Programme FP7/248942 (2009–2013).

[2] The interdisciplinary Transregional Collaborative Research Center Spatial Cognition: Reasoning, Action, Interaction has been established by the German Research Foundation (DFG) (2003–2014).

First-MM project[3] aims to enable robots to acquire new manipulation skills which also involve grasping and operating articulated objects using our approach. The goal of the TidyUpRobot project[4] is to use the PR2 robot in various tidying-up tasks. As part of this project, the PR2 robot will have to annotate floor plans with articulated objects and their respective kinematic models.

To conclude, we think that mobile manipulation robots have a large application potential in the near future. In this work, we presented several innovative approaches to relevant problems that appear when mobile manipulators are applied in unstructured environments and changing situations. We hope that our work increases the dependability, flexibility, and ease of use of manipulation robots and thereby contributes to the development of truly useful robotic assistants for industry and society.

[3] First-MM is another research project founded under the European Union Seventh Framework Programme FP7/248258 (2010–2014).

[4] The TidyUpRobot project is part of the PR2 beta program sponsored by Willow Garage (2010–2012).

Appendix A
The Laplace Approximation

The Laplace approximation is a technique that can be used to approximate the integral $\int f(\mathbf{x})\ \mathrm{d}\mathbf{x}$ of a function $f(\mathbf{x})$. The general idea is to approximate the normalized probability distribution $p(\mathbf{x}) = f(\mathbf{x})/Z$ with a multivariate Gaussian distribution and to use this approximation for estimating the normalization constant $Z = \int f(\mathbf{x})\ \mathrm{d}\mathbf{x}$. In the following, we summarize the derivation of the Laplace approximation according to the text book of Bishop (2007). We introduce the Laplace approximation here because we need it later as a tool during the derivation of the BIC in Appendix B.

Given a function $f(\mathbf{x})$, the Laplace approximation assumes that the function has a peak at $\hat{\mathbf{x}}$. At this point, we Taylor-expand the log density function around $\hat{\mathbf{x}}$ to the second order which gives us

$$\begin{aligned}\log f(\mathbf{x}) \simeq& \log f(\hat{\mathbf{x}}) \\ &+ \nabla \log f(\mathbf{x})\big|_{\mathbf{x}=\hat{\mathbf{x}}}(\mathbf{x}-\hat{\mathbf{x}}) \\ &+ \frac{1}{2}(\mathbf{x}-\hat{\mathbf{x}})^T \left(\nabla\nabla \log f(\mathbf{x})\big|_{\mathbf{x}=\hat{\mathbf{x}}}\right)(\mathbf{x}-\hat{\mathbf{x}}),\end{aligned} \tag{A.1}$$

where ∇ is the gradient operator. As the point $\hat{\mathbf{x}}$ is at a maximum of $f(\mathbf{x})$ and, consequently, of $\log f(\mathbf{x})$, the first-order derivative $\nabla \log f(\mathbf{x})\big|_{\mathbf{x}=\hat{\mathbf{x}}}$ equals zero. As a result the second term in Eq. (A.1) vanishes. Further, by denoting the negative of the second-order derivative as

$$A = -\nabla\nabla \log f(\mathbf{x})\big|_{\mathbf{x}=\hat{\mathbf{x}}}, \tag{A.2}$$

we can rewrite Eq. (A.1) as

$$\log f(\mathbf{x}) \simeq \log f(\hat{\mathbf{x}}) - \frac{1}{2}(\mathbf{x}-\hat{\mathbf{x}})^T A(\mathbf{x}-\hat{\mathbf{x}}), \tag{A.3}$$

Taking the exponential on both sides we obtain

$$f(\mathbf{x}) \simeq f(\hat{\mathbf{x}}) \exp\left(-\frac{1}{2}(\mathbf{x}-\hat{\mathbf{x}})^T A(\mathbf{x}-\hat{\mathbf{x}})\right). \tag{A.4}$$

Remember that a multivariate, k-dimensional Gaussian distribution has a (normalized) probability density function of

$$\mathcal{N}(\mathbf{x};\boldsymbol{\mu},\Sigma) = \frac{1}{(2\pi)^{k/2}} \frac{1}{|\Sigma|^{1/2}} \exp\left(-\frac{1}{2}(\mathbf{x}-\boldsymbol{\mu})^T\Sigma^{-1}(\mathbf{x}-\boldsymbol{\mu})\right). \tag{A.5}$$

Identifying the parameters with $\boldsymbol{\mu} = \hat{\mathbf{x}}$ and $\Sigma = A^{-1}$ we obtain a Gaussian approximation $q(\mathbf{x})$ of $p(\mathbf{x})$ with

$$p(\mathbf{x}) \simeq q(\mathbf{x}) := \frac{|A|^{1/2}}{(2\pi)^{k/2}} \exp\left(-\frac{1}{2}(\mathbf{x} - \hat{\mathbf{x}})^T A(\mathbf{x} - \hat{\mathbf{x}})\right) \tag{A.6}$$

Comparing Eq. (A.4) with Eq. (A.6) yields (together with $p(\mathbf{x}) = f(\mathbf{x})/Z$) an approximation of the normalization constant Z, i.e.,

$$\frac{f(\hat{\mathbf{x}})}{Z} \simeq \frac{|A|^{1/2}}{(2\pi)^{k/2}}. \tag{A.7}$$

When we solve this equation for the normalization constant Z, we obtain an estimate of the integral $\int f(\mathbf{x})\,\mathrm{d}\mathbf{x}$, i.e.,

$$Z = \int f(\mathbf{x})\,\mathrm{d}\mathbf{x} \simeq \frac{(2\pi)^{k/2}}{|A|^{1/2}} f(\hat{\mathbf{x}}) \tag{A.8}$$

which we will use in Appendix B for deriving the Bayesian information criterion.

Appendix B
Derivation of the Bayesian Information Criterion

In this section, we derive the Bayesian Information Criterion (BIC). The BIC is an approximation of the model evidence $p(\mathcal{D} \mid \mathcal{M})$ for choosing between alternative models. The BIC was first proposed by Schwarz (1978). In this appendix, we derive the BIC ourselves step by step. The derivation is based on the book of Bishop (2007). Our goal is to provide the reader with a thorough understanding of the BIC including its underlying assumptions as needed for Chapter 4.

The BIC assumes that the prior over the parameters is broad, i.e., $p(\boldsymbol{\theta}) = \mathcal{N}(\boldsymbol{\theta}; \boldsymbol{\mu}, \Sigma)$ is a Gaussian distribution that has a covariance matrix Σ with a large determinant. Further, the BIC assumes that the data samples in $\mathcal{D} = \{(\mathbf{x}_i)\}_{i=1}^{n}$ are independent and identically distributed (i.i.d.).

Remember from Section 2.2.4 that the model evidence can be computed by integrating over the whole parameter space of a model, i.e.,

$$p(\mathcal{D} \mid \mathcal{M}) = \int p(\mathcal{D} \mid \mathcal{M}, \boldsymbol{\theta}) p(\boldsymbol{\theta} \mid \mathcal{M}) \, \mathrm{d}\boldsymbol{\theta}. \tag{B.1}$$

The BIC assumes that the posterior $p(\boldsymbol{\theta} \mid \mathcal{M}, \mathcal{D}) \propto p(\mathcal{D} \mid \mathcal{M}, \boldsymbol{\theta}) p(\boldsymbol{\theta} \mid \mathcal{M})$ has a strong peak at the maximum-a-posteriori parameter vector when $\mathcal{D}$ is fixed, i.e.,

$$\hat{\boldsymbol{\theta}} = \arg\max_{\boldsymbol{\theta}} p(\mathcal{D} \mid \mathcal{M}, \boldsymbol{\theta}) p(\boldsymbol{\theta} \mid \mathcal{M}), \tag{B.2}$$

so that it can be approximated with a Gaussian. This is accomplished by using Laplace's approximation as described in Appendix A: we identify $f(\boldsymbol{\theta}) = p(\mathcal{D} \mid \mathcal{M}, \boldsymbol{\theta}) p(\boldsymbol{\theta} \mid \mathcal{M})$ and use Eq. (A.8) for approximating the integral $Z = \int f(\boldsymbol{\theta}) \, \mathrm{d}\boldsymbol{\theta}$. This gives us

$$p(\mathcal{D} \mid \mathcal{M}) \simeq \frac{(2\pi)^{k/2}}{|A|^{1/2}} p(\mathcal{D} \mid \mathcal{M}, \hat{\boldsymbol{\theta}}) p(\hat{\boldsymbol{\theta}} \mid \mathcal{M}), \tag{B.3}$$

where $A = -\nabla\nabla \log p(\mathcal{D} \mid \mathcal{M}, \boldsymbol{\theta}) p(\boldsymbol{\theta} \mid \mathcal{M})\big|_{\boldsymbol{\theta}=\hat{\boldsymbol{\theta}}}$ is the Hessian of the posterior probability distribution evaluated at $\hat{\boldsymbol{\theta}}$. By taking the log on both sides, we obtain

$$\log p(\mathcal{D} \mid \mathcal{M}) \simeq \log p(\mathcal{D} \mid \mathcal{M}, \hat{\boldsymbol{\theta}}) + \log p(\hat{\boldsymbol{\theta}} \mid \mathcal{M}) + \frac{k}{2} \log(2\pi) - \frac{1}{2} \log |A|, \tag{B.4}$$

The Hessian can be split accordingly into two components relating to the data likelihood and the prior over the parameter space, i.e.,

$$
\begin{aligned}
A &= -\nabla\nabla \log p(\mathcal{D} \mid \mathcal{M}, \boldsymbol{\theta}) p(\boldsymbol{\theta} \mid \mathcal{M})\big|_{\boldsymbol{\theta}=\hat{\boldsymbol{\theta}}} && \text{(B.5)}\\
&= -\nabla\nabla \log p(\mathcal{D} \mid \mathcal{M}, \boldsymbol{\theta})\big|_{\boldsymbol{\theta}=\hat{\boldsymbol{\theta}}} - \nabla\nabla \log p(\boldsymbol{\theta} \mid \mathcal{M})\big|_{\boldsymbol{\theta}=\hat{\boldsymbol{\theta}}} && \text{(B.6)}\\
&= H + \Sigma^{-1}, && \text{(B.7)}
\end{aligned}
$$

where H is the matrix of second derivatives of the negative log likelihood $p(\mathcal{D} \mid \mathcal{M}, \boldsymbol{\theta})$ evaluated at $\hat{\boldsymbol{\theta}}$. If the prior over the parameters is broad, or the number of data samples in $\mathcal{D}$ is large, the Hessian will be dominated mostly by the first term and Σ^{-1} can be neglected so that we approximate

$$A \simeq H. \tag{B.8}$$

Together with our assumption that the parameters are normally distributed, i.e.,

$$\log p(\boldsymbol{\theta} \mid \mathcal{M}) = \frac{1}{2}(\boldsymbol{\theta} - \boldsymbol{\mu})^T \Sigma^{-1} (\boldsymbol{\theta} - \boldsymbol{\mu}) + \text{const}, \tag{B.9}$$

we can rewrite Eq. (B.4) using Eq. (B.8) and Eq. (B.9) as

$$
\begin{aligned}
&\log p(\mathcal{D} \mid \mathcal{M}) \simeq \\
&\quad \log p(\mathcal{D} \mid \mathcal{M}, \hat{\boldsymbol{\theta}}) + \frac{1}{2}(\hat{\boldsymbol{\theta}} - \boldsymbol{\mu})^T \Sigma^{-1} (\hat{\boldsymbol{\theta}} - \boldsymbol{\mu}) - \frac{1}{2}\log|H| + \text{const}. && \text{(B.10)}
\end{aligned}
$$

By using again the assumption that the prior over the parameters is broad, Σ^{-1} is small and the second term on the right-hand side in Eq. (B.10) vanishes. Note that the same happens if the number of data samples is sufficiently large: as the prior is constant in the number of data samples, the data likelihood term dominates over the prior for large n. As an intermediate result, we approximate the model evidence as

$$\log p(\mathcal{D} \mid \mathcal{M}) \simeq \log p(\mathcal{D} \mid \mathcal{M}, \hat{\boldsymbol{\theta}}) - \frac{1}{2}\log|H| + \text{const}. \tag{B.11}$$

In the next step of the BIC approximation, we consider the second term in Eq. (B.11) regarding the Hessian H of the data likelihood. When the data samples in $\mathcal{D}$ are i.i.d., then the Hessian corresponds to the sum of Hessians induced by the individual data samples. This means that we can factorize the Hessian, i.e.,

$$
\begin{aligned}
H &= -\nabla\nabla \log p(\mathcal{D} \mid \mathcal{M}, \boldsymbol{\theta})\big|_{\boldsymbol{\theta}=\hat{\boldsymbol{\theta}}} && \text{(B.12)}\\
&= -\nabla\nabla \log \prod_{i=1,\ldots,n} p(\mathbf{x}_i \mid \mathcal{M}, \boldsymbol{\theta})\big|_{\boldsymbol{\theta}=\hat{\boldsymbol{\theta}}} && \text{(B.13)}\\
&= \sum_{i=1,\ldots,n} -\nabla\nabla \log p(\mathbf{x}_i \mid \mathcal{M}, \boldsymbol{\theta})\big|_{\boldsymbol{\theta}=\hat{\boldsymbol{\theta}}} && \text{(B.14)}
\end{aligned}
$$

By denoting the contribution of a single data sample to the Hessian as

$$H_i = -\nabla\nabla \log p(\mathbf{y}_i \mid \mathbf{x}_i, \mathcal{M}, \boldsymbol{\theta})\big|_{\boldsymbol{\theta}=\hat{\boldsymbol{\theta}}}, \tag{B.15}$$

we can now express the Hessian as the sum of the individual contributions, i.e.,

$$H = \sum_{i=1,\dots,n} H_i = n\hat{H}. \tag{B.16}$$

and finally approximate it as the product of the number of data samples n and the average sample contribution $\hat{H}$. Using this decomposition, we can now approximate the determinant of H in the third term in Eq. (B.10) using

$$\log|H| = \log|n\hat{H}| \tag{B.17}$$

$$= \log\left(n^k|\hat{H}|\right) \tag{B.18}$$

$$= k\log n + \log|\hat{H}|, \tag{B.19}$$

where k corresponds to the number of parameters, i.e., $\boldsymbol{\theta} \in \mathbb{R}^k$. Note that we assume for this approximation that the sample-wise average Hessian $\hat{H}$ has full rank. Combining this result with Eq. (B.11), we obtain an approximation of the model evidence that is exclusively based on the data likelihood $p(\mathcal{D} \mid \mathcal{M}, \hat{\boldsymbol{\theta}})$, the number of model parameters k and the number of data samples n.

$$\log p(\mathcal{D} \mid \mathcal{M}) \simeq \log p(\mathcal{D} \mid \mathcal{M}, \hat{\boldsymbol{\theta}}) - \frac{k}{2}\log n + \text{const.} \tag{B.20}$$

The Bayesian information criterion is typically defined as the negative logarithm of the model evidence multiplied by two, i.e.,

$$\text{BIC} = -2\log p(\mathcal{D} \mid \mathcal{M}) \tag{B.21}$$

$$\simeq -2\log p(\mathcal{D} \mid \mathcal{M}, \hat{\boldsymbol{\theta}}) + k\log n + \text{const}, \tag{B.22}$$

so that the model with the lowest BIC score is the one to be preferred.

References

Abbeel, P., Coates, A., Quigley, M., Ng, A.Y.: An application of reinforcement learning to aerobatic helicopter flight. In: Proc. of the Conf. on Neural Information Processing Systems (NIPS), Vancouver, Canada (2007)

Agarwal, S., Awan, A., Roth, D.: Learning to detect objects in images via a sparse, part-based representation. IEEE Transactions on Pattern Analysis and Machine Intelligence 26(11), 1475–1490 (2004)

Allen, P.K.: Integrating vision and touch for object recognition tasks. Intl. Journal of Robotics Research (IJRR) 7(6), 15–33 (1988)

Anderson, D., Herman, H., Kelly, A.: Experimental characterization of commercial flash ladar devices. In: Proc. of the Intl. Conf. on Sensing and Technology, Palmerston North, New Zealand (2005)

Andreopoulos, A., Tsotsos, J.K.: Active vision for door localization and door opening using playbot. In: Proc. of the Canadian Conf. on Computer and Robot Vision (CRV), Windsor, Canada (2008)

Anguelov, D., Koller, D., Parker, E., Thrun, S.: Detecting and modeling doors with mobile robots. In: Proc. of the IEEE Intl. Conf. on Robotics and Automation (ICRA), New Orleans, LA, USA (2004)

de Angulo, V.R., Torras, C.: Using pSOMs to learn inverse kinematics through virtual decomposition of the robot. In: Cabestany, J., Prieto, A.G., Sandoval, F. (eds.) IWANN 2005. LNCS, vol. 3512, pp. 701–708. Springer, Heidelberg (2005)

Argall, B.D., Chernova, S., Veloso, M., Browning, B.: A survey of robot learning from demonstration. Robotics and Autonomous Systems (RAS) 57(5), 469–483 (2009)

Asfour, T., Gyarfas, F., Azad, P., Dillmann, R.: Imitation learning of dual-arm manipulation tasks in humanoid robots. In: Proc. of the IEEE-RAS Intl. Conf. on Humanoid Robots (Humanoids), Genova, Italy (2006)

Bakker, P., Kuniyoshi, Y.: Robot see, robot do: An overview of robot imitation. In: Proc. of the Workshop on Learning in Robots and Animals (AISB), Sussex, UK (1996)

Beetz, M., Jain, D., Mösenlechner, L., Tenorth, M.: Towards performing everyday manipulation activities. Robotics and Autonomous Systems (RAS) 58(9), 1085–1095 (2010)

Bentivegna, D.C., Atkeson, C.G., Cheng, G.: Learning tasks from observation and practice. Robotics and Autonomous Systems (RAS) 47(2-3), 163–169 (2004)

Bicchi, A., Kumar, V.: Robotic grasping and manipulation. In: Proc. of the IEEE Intl. Conf. on Robotics and Automation, ICRA (2001)

Bierbaum, A., Rambow, M., Asfour, T., Dillmann, R.: Grasp affordances from multi-fingered tactile exploration using dynamic potential fields. In: Proc. of the IEEE-RAS Intl. Conf. on Humanoid Robots (Humanoids), Paris, France (2009)

Billard, A., Calinon, S., Dillmann, R., Schaal, S.: Robot programming by demonstration. In. In: Siciliano, B., Khatib, O. (eds.) Handbook of Robotics, ch. 59. Springer (2008)

Bishop, C.M.: Pattern Recognition and Machine Learning. Information Science and Statistics. Springer (2007)

Bongard, J., Lipson, H.: Automated reverse engineering of nonlinear dynamical systems. Proc. of the National Academy of Sciences 104(24), 9943–9948 (2007)

Bongard, J., Zykov, V., Lipson, H.: Resilient machines through continuous self-modeling. Science 314(5802), 1118–1121 (2006a)

Bongard, J., Zykov, V., Lipson, H.: Automated synthesis of body schema using multiple sensor modalities. In: Proc. of the Intl. Conf. on the Simulation and Synthesis of Living Systems, Bloomington, IN, USA (2006b)

Bradski, G., Kaehler, A.: Learning OpenCV: Computer Vision with the OpenCV Library. O'Reilly Media (2008)

Bromberg, F., Margaritis, D., Honavar, V.: Efficient Markov network structure discovery using independence tests. Journal of Artificial Intelligence Research (JAIR) 35(1), 449–485 (2009)

Brox, T., Rosenhahn, B., Gall, J., Cremers, D.: Combined region- and motion-based 3D tracking of rigid and articulated objects. IEEE Transactions on Pattern Analysis and Machine Intelligence 32(2), 402–415 (2010)

Buss, S.R., Kim, J.: Selectively damped least squares for inverse kinematics. Journal of Graphics Tools 10(3), 37–49 (2005)

Calinon, S., Billard, A.: A probabilistic programming by demonstration framework handling skill constraints in joint space and task space. In: Proc. of the IEEE/RSJ Intl. Conf. on Intelligent Robots and Systems (IROS), Nice, France (2008)

Calinon, S., Billard, A.: Statistical learning by imitation of competing constraints in joint space and task space. Advanced Robotics 23(15), 2059–2076 (2009)

Calinon, S., Guenter, F., Billard, A.: Goal-directed imitation in a humanoid robot. In: Proc. of the IEEE Intl. Conf. on Robotics and Automation (ICRA), Barcelona, Spain (2005)

Castelli, F.: An integrated tactile-thermal robot sensor with capacitive tactile array. IEEE Transactions on Industry Applications 38(1), 85–90 (2002)

Chalodhorn, R., Grimes, D.B., Rao, R.P.N.: Learning to walk through imitation. In: Proc. of the Intl. Conf. on Artificial Intelligence (IJCAI), San Mateo, CA, USA (2007)

Chickering, D.M.: Learning Bayesian networks is NP-Complete. In: Fisher, D., Lenz, H. (eds.) Learning from Data: Artificial Intelligence and Statistics, pp. 121–130. Springer (1996)

Chickering, D.M.: Learning equivalence classes of Bayesian-network structures. Journal of Machine Learning Research (JMLR) 2, 445–498 (2002)

Chitta, S., Cohen, B., Likhachev, M.: Planning for autonomous door opening with a mobile manipulator. In: Proc. of the IEEE Intl. Conf. on Robotics and Automation (ICRA), Anchorage, AK, USA (2010)

Chuang, C., Chen, R.: 3D capacitive tactile sensor using DRIE micromachining. In: Smart Sensors, Actuators, and MEMS II, Seville, Spain (2005)

Chum, O., Matas, J.: Matching with PROSAC - progressive sample consensus. In: Proc. of the IEEE Conf. on Computer Vision and Pattern Recognition (CVPR), San Diego, CA, USA (2005)

Ciocarlie, M., Allen, P.: Data-driven optimization for underactuated robotic hands. In: Proc. of the IEEE Intl. Conf. on Robotics and Automation, ICRA (2010)

Coates, A., Abbeel, P., Ng, A.Y.: Learning for control from multiple demonstrations. In: Proc. of the Intl. Conf. on Machine Learning (ICML), Helsinki, Finland (2008)

Collet Romea, A., Martinez Torres, M., Srinivasa, S.: The MOPED framework: Object recognition and pose estimation for manipulation. Intl. Journal of Robotics Research (IJRR) 30(10) (2011)

Cormen, T.H., Leiserson, C.E., Rivest, R.L., Stein, C.: Introduction to Algorithms. MIT Press (2001)

Craig, J.J.: Introduction to Robotics: Mechanics and Control. Addison-Wesley Publishing Company (1989)

Csurka, G., Dance, L., Willamowski, J., Bray, C.: Visual categorization with bags of keypoints. In: Proc. of the Workshop on Statistical Learning at the Europ. Conf. on Computer Vision (ECCV), Prague, Czech Republic (2004)

Curless, B., Levoy, M.: Better optical triangulation through spacetime analysis. In: Proc. of the Intl. Conf. on Computer Vision (ICCV), Boston, MA, USA (1995)

Dahiya, R.S., Metta, G., Valle, M., Sandini, G.: Tactile sensing: From humans to humanoids. IEEE Transactions on Robotics (T-RO) 26(1), 1–20 (2010)

Daly, R., Shen, Q.: Learning Bayesian network equivalence classes with ant colony optimization. Journal of Artificial Intelligence Research (JAIR) 35(1), 391–447 (2009)

Dearden, A., Demiris, Y.: Learning forward models for robots. In: Proc. of the Intl. Conf. on Artificial Intelligence (IJCAI), Edinburgh, Scotland (2005)

Dellaert, F.: Square root SAM. In: Proc. of Robotics: Science and Systems (RSS), Cambridge, MA, USA (2005)

Dempster, A.P., Laird, N.M., Rubin, D.B.: Maximum likelihood from incomplete data via the EM algorithm. Journal of the Royal Statistical Society, Series B 39(1), 1–38 (1977)

Diankov, R., Srinivasa, S., Ferguson, D., Kuffner, J.: Manipulation planning with caging grasps. In: Proc. of the IEEE-RAS Intl. Conf. on Humanoid Robots (Humanoids), Daejeon, Korea (2008)

Dollar, A.M., Howe, R.D.: The highly adaptive SDM hand: Design and performance evaluation. Intl. Journal of Robotics Research (IJRR) 29(5) (2010)

Doulgeri, Z., Arimoto, S.: Force position control for a robot finger with a soft tip and kinematic uncertainties. Robotics and Autonomous Systems (RAS) 55(4), 328–336 (2007)

D'Souza, A., Vijayakumar, S., Schaal, S.: Learning inverse kinematics. In: Proc. of the IEEE/RSJ Intl. Conf. on Intelligent Robots and Systems (IROS), Maui, HI, USA (2001)

Duda, R.O., Hart, P.E., Stork, D.G.: Pattern Classification and Scene Analysis. John Wiley & Sons Inc. (1973)

Eppner, C.: Techniques for the imitation of manipulative actions by robots. MSc thesis, University of Freiburg, Germany (2008)

Featherstone, R., Orin, D.: Dynamics. In: Siciliano, B., Khatib, O. (eds.) Handbook of Robotics, ch. 2. Springer (2008)

Fei-Fei, L., Perona, P.: A Bayesian hierarchical model for learning natural scene categories. In: Proc. of the IEEE Conf. on Computer Vision and Pattern Recognition (CVPR), San Diego, CA, USA (2005)

Fiala, M.: ARtag, a fiducial marker system using digital techniques. In: Proc. of the IEEE Conf. on Computer Vision and Pattern Recognition (CVPR), San Diego, CA, USA (2005)

Fischler, M., Bolles, R.: Random sample consensus: a paradigm for model fitting with application to image analysis and automated cartography. Communications of the ACM 24(6), 381–395 (1981)

Fox, D., Ren, X.: Overview of RGB-D cameras and open research issues. In: Proc. of the Workshop on Advanced Reasoning with Depth Cameras at Robotics: Science and Systems (RSS), Zaragoza, Spain (2010)

Frank, B., Schmedding, R., Stachniss, C., Teschner, M., Burgard, W.: Learning the elasticity parameters of deformable objects with a manipulation robot. In: Proc. of the IEEE/RSJ Intl. Conf. on Intelligent Robots and Systems (IROS), Taipei, Taiwan (2010)

Frese, U.: Treemap: An $O(\log n)$ algorithm for indoor simultaneous localization and mapping. Autonomous Robots 21(2), 103–122 (2006)

Gallagher, S.: How the Body Shapes the Mind. Oxford University Press, USA (2005)

Gaskett, C., Cheng, G.: Online learning of a motor map for humanoid robot reaching. In: Proc. of the Intl. Conf. on Computational Intelligence, Robotics and Autonomous Systems (CIRAS), Singapore (2003)

Gatla, C.S., Lumia, R., Wood, J., Starr, G.: An automated method to calibrate industrial robots using a virtual closed kinematic chain. IEEE Transactions on Robotics (T-RO) 23(6), 1105–1116 (2007)

Gibson, J.J.: The theory of affordances. In: Shaw, R., Bransford, J. (eds.) Perceiving, Acting, and Knowing: Toward an Ecological Psychology, pp. 67–82. Lawrence Erlbaum (1977)

Gordon, I., Lowe, D.G.: What and where: 3D object recognition with accurate pose. In: Ponce, J., Hebert, M., Schmid, C., Zisserman, A. (eds.) Toward Category-Level Object Recognition. LNCS, vol. 4170, pp. 67–82. Springer, Heidelberg (2006)

Gorges, N., Navarro, S.E., Göger, D., Wörn, H.: Haptic object recognition using passive joints and haptic key features. In: Proc. of the IEEE Intl. Conf. on Robotics and Automation (ICRA), Anchorage, AK, USA (2010)

Grebenstein, M., Albu-Schäffer, A., Bahls, T., Chalon, M., Eiberger, O., Friedl, W., Gruber, R., Haddadin, S., Hagn, U., Haslinger, R., Hoppner, H., Jörg, S., Nickl, M., Nothhelfer, A., Petit, F., Reill, J., Seitz, N., Wimböck, T., Wolf, S., Wüsthoff, T., Hirzinger, G.: The dlr hand arm system. In: Proc. of the IEEE Intl. Conf. on Robotics and Automation, ICRA (2011)

Grimes, D., Chalodhorn, R., Rao, R.: Dynamic imitation in a humanoid robot through nonparametric probabilistic inference. In: Proc. of Robotics: Science and Systems (RSS), Philadelphia, PA, USA (2006)

Grisetti, G., Stachniss, C., Burgard, W.: Non-linear constraint network optimization for efficient map learning. IEEE Transactions on Intelligent Transportation systems 10(3), 428–439 (2009)

Grisetti, G., Kümmerle, R., Stachniss, C., Frese, U., Hertzberg, C.: Hierarchical optimization on manifolds for online 2D and 3D mapping. In: Proc. of the IEEE Intl. Conf. on Robotics and Automation (ICRA), Anchorage, AK, USA (2010)

Grochow, K., Martin, S.L., Hertzmann, A., Popović, Z.: Style-based inverse kinematics. ACM Transactions on Graphics (TOG) 23(3), 522–531 (2004)

Gupta, M., Sukhatme, G.: Using manipulation primitives for brick sorting in clutter. In: Proc. of the IEEE Intl. Conf. on Robotics and Automation, ICRA (2012)

Hafner, R., Riedmiller, M.: Neural reinforcement learning controllers for a real robot application. In: Proc. of the IEEE Intl. Conf. on Robotics and Automation (ICRA), Rome, Italy (2007)

Hasegawa, Y., Shikida, M., Shimizu, T., Miyaji, T., Sasaki, H., Sato, K., Itoigawa, K.: Micromachined active tactile sensor for hardness detection. Sensors and Actuators A: Physical 114(2-3), 141–146 (2004)

Hausman, K., Balint-Benczedi, F., Pangercic, D., Marton, Z.-C., Ueda, R., Okada, K., Beetz, M.: Tracking-based interactive segmentation of textureless objects. In: Proc. of the IEEE Intl. Conf. on Robotics and Automation, ICRA (2013)

He, R., Zhao, Y., Yang, S., Yang, S.: Kinematic-parameter identification for serial-robot calibration based on POE formula. IEEE Transactions on Robotics (T-RO) 26(3), 411–423 (2010)

Hersch, M., Sauser, E., Billard, A.: Online learning of the body schema. Intl. Journal of Humanoid Robotics 5(2), 161–181 (2008)

Hetzel, G., Leibe, B., Levi, P., Schiele, B.: 3D object recognition from range images using local feature histograms. In: Proc. of the IEEE Conf. on Computer Vision and Pattern Recognition (CVPR), Kauai, HI, USA (2001)

Hinterstoisser, S., Lepetit, V., Ilic, S., Holzer, S., Bradski, G., Konolige, K., Navab, N.: Model based training, detection and pose estimation of texture-less 3D objects in heavily cluttered scenes. In: Proc. of the Asian Conference on Computer Vision, ACCV (2012)

Hinterstoisser, S., Holzer, S., Cagniart, C., Ilic, S., Konolige, K., Navab, N., Lepetit, V.: Multimodal templates for real-time detection of texture-less objects in heavily cluttered scenes. In: Proc. of the Intl. Conf. on Computer Vision, ICCV (2011)

Hoffmann, M., Marques, H., Hernandez Arieta, A., Sumioka, H., Lungarella, M., Pfeifer, R.: Body schema in robotics: a review. IEEE Transactions on Autonomous Mental Development 2(4), 304–324 (2010)

Hollerbach, J., Wampler, C.W.: The calibration index and taxonomy for robot kinematic calibration methods. Intl. Journal of Robotics Research (IJRR) 15(6) (1996)

Hollerbach, J., Khalil, W., Gautier, M.: Model identification. In: Siciliano, B., Khatib, O. (eds.) Handbook of Robotics, ch. 14. Springer (2008)

Ijspeert, A.J., Nakanishi, J., Schaal, S.: Learning attractor landscapes for learning motor primitives. In: Proc. of the Conf. on Neural Information Processing Systems (NIPS), Vancouver, Canada (2002)

Jain, A., Kemp, C.C.: Behavior-based door opening with equilibrium point control. In: Proc. of the Workshop on Mobile Manipulation in Human Environments at Robotics: Science and Systems (RSS), Seattle, WA, USA (2009a)

Jain, A., Kemp, C.C.: Pulling open novel doors and drawers with equilibrium point control. In: Proc. of the IEEE-RAS Intl. Conf. on Humanoid Robots (Humanoids), Paris, France (2009b)

Jain, A., Kemp, C.C.: Pulling open doors and drawers: Coordinating an omni-directional base and a compliant arm with equilibrium point control. In: Proc. of the IEEE Intl. Conf. on Robotics and Automation (ICRA), Anchorage, AK, USA (2010)

Jain, A., Nguyen, H., Rath, M., Okerman, J., Kemp, C.C.: The Complex Structure of Simple Devices: A Survey of Trajectories and Forces that Open Doors and Drawers. In: Proc. of the IEEE RAS/EMBS Intl. Conf. on Biomedical Robotics and Biomechatronics, BIOROB (2010)

Jeannerod, M.: Object oriented action. In: Bennet, K.M.B., Castiello, U. (eds.) Insights into the Reach to Grasp Movement, pp. 3–15. North-Holland (2007)

Jensen, F.V.: Bayesian Networks and Decision Graphs. Springer (2001)

Johansson, R.S.: Sensory and memory information in the control of dexterous manipulation. In: Lacquaniti, F., Viviani, P. (eds.) Neural Bases of Motor Behaviour, pp. 205–260. Kluwer Academic Publishers (1996)

Johansson, R.S., Flanagan, J.R.: Coding and use of tactile signals from the fingertips in object manipulation tasks. Nature Reviews Neuroscience 10, 345–359 (2009)

Katz, D., Brock, O.: Manipulating articulated objects with interactive perception. In: Proc. of Robotics: Science and Systems (RSS), Pasadena, CA, USA (2008)

Kemp, C.C.: A wearable system that learns a kinematic model and finds structure in everyday manipulation by using absolute orientation sensors and a camera. PhD thesis, Massachusetts Institute of Technology, Cambridge, MA, USA (2005)

Keogh, E.J., Pazzani, M.J.: Derivative dynamic time warping. In: Proc. of the SIAM Intl. Conf. on Data Mining (SDM), Chicago, IL, USA (2001)

Kirk, A., O'Brien, J.F., Forsyth, D.A.: Skeletal parameter estimation from optical motion capture data. In: Proc. of the Intl. Conf. on Computer Graphics and Interactive Techniques (SIGGRAPH), Los Angeles, CA, USA (2004)

Klingbeil, E., Saxena, A., Ng, A.Y.: Learning to open new doors. In: Proc. of the Workshop on Robot Manipulation at Robotics: Science and Systems (RSS), Seattle, WA, USA (2009)

Kober, J., Peters, J.: Learning motor primitives for robotics. In: Proc. of the IEEE Intl. Conf. on Robotics and Automation (ICRA), Kobe, Japan (2009)

Koller, D., Friedman, N.: Probabilistic Graphical Models: Principles and Techniques. MIT Press (2009)

Kolter, J., Ng, A.: Learning omnidirectional path following using dimensionality reduction. In: Proc. of Robotics: Science and Systems (RSS), Atlanta, GA, USA (2007)

Konolige, K.: Small vision systems: hardware and implementation. In: Proc. of the Intl. Symp. on Robotics Research, Hayama, Japan (1997)

Konolige, K.: Projected texture stereo. In: Proc. of the IEEE Intl. Conf. on Robotics and Automation (ICRA), Anchorage, AK, USA (2010)

Kormushev, P., Calinon, S., Saegusa, R., Metta, G.: Learning the skill of archery by a humanoid robot iCub. In: Proc. of the IEEE-RAS Intl. Conf. on Humanoid Robots (Humanoids), Nashville, TN, USA (2010)

Kormushev, P., Calinon, S., Caldwell, D.G.: Imitation learning of positional and force skills demonstrated via kinesthetic teaching and haptic input. In: Advanced Robotics (2011) (to appear)

Kragic, D., Petersson, L., Christensen, H.I.: Visually guided manipulation tasks. Robotics and Autonomous Systems (RAS) 40(2-3), 193–203 (2002)

Krainin, M., Henry, P., Ren, X., Fox, D.: Manipulator and object tracking for in hand model acquisition. In: Proc. of the IEEE Intl. Conf. on Robotics and Automation (ICRA), Anchorage, AK, USA (2010)

Kuipers, B., Byun, Y.-T.: A robust, qualitative method for robot spatial learning. In: Proc. of the National Conf. on Artificial Intelligence (AAAI), Saint Paul, MN, USA (1988)

Kuipers, B., Browning, R., Gribble, B., Hewett, M., Remolina, E.: The spatial semantic hierarchy. Artificial Intelligence 119, 191–233 (2000)

Kumar, S., Behera, L., McGinnity, T.M.: Kinematic control of a redundant manipulator using an inverse-forward adaptive scheme with a KSOM based hint generator. Robotics and Autonomous Systems (RAS) 58(5), 622–633 (2010)

Kuss, M.: Gaussian process models for robust regression, classification, and reinforcement learning. PhD thesis, University of Darmstadt, Germany (2008)

LaValle, S.M.: Planning Algorithms. Cambridge University Press (2006)

Lawrence, N.D.: Probabilistic non-linear principal component analysis with Gaussian process latent variable models. Journal of Machine Learning Research (JMLR) 6, 1783–1816 (2005)

Lee, M.H., Nicholls, H.R.: Tactile sensing for mechatronics – a state of the art survey. Mechatronics 9(1), 1–31 (1999)

Lewis, D.D.: Naive (Bayes) at forty: The independence assumption in information retrieval. In: Nédellec, C., Rouveirol, C. (eds.) ECML 1998. LNCS, vol. 1398, pp. 4–15. Springer, Heidelberg (1998)

Lim, J.: Optimized projection pattern supplementing stereo systems. In: Proc. of the IEEE Intl. Conf. on Robotics and Automation (ICRA), Kobe, Japan (2009)

Lopes, M., Santos-Victor, J.: Visual learning by imitation with motor representations. IEEE Transactions on Systems, Man, and Cybernetics, Part B: Cybernetics 35(3), 438–449 (2005)

Lu, F., Milios, E.: Globally consistent range scan alignment for environment mapping. Autonomous Robots 4(4), 333–349 (1997)

Ma, Y., Soatto, S., Kosecká, J., Sastry, S.S.: An Invitation to 3-D Vision: From Images to Geometric Models. In: Interdisciplinary Applied Mathematics: Imaging, Vision, and Graphics. Springer (2003)

Machline, M., Arieli, Y., Sphpunt, A., Freedman, B.: Depth mapping using projected patterns, US patent 20100118123. Prime Sense Ltd. (2010)

MacKay, D.J.C.: Information Theory, Inference, and Learning Algorithms. Cambridge University Press (2003)

Maeno, T.: Friction estimation by pressing an elastic finger-shaped sensor against a surface. IEEE Transactions on Robotics and Automation 20(2), 222–228 (2004)

Maravita, A., Iriki, A.: Tools for the body (schema). Trends in Cognitive Sciences 8(2), 79–86 (2004)

Margaritis, D., Thrun, S.: Bayesian network induction via local neighborhoods. In: Proc. of the Conf. on Neural Information Processing Systems (NIPS), Denver, CO, USA (1999)

Martinez-Cantin, R., Lopes, M., Montesano, L.: Body schema acquisition through active learning. In: Proc. of the IEEE Intl. Conf. on Robotics and Automation (ICRA), Anchorage, AK, USA (2010)

Mason, M.T., Rodriguez, A., Srinivasa, S., Vazquez, S.S.: Autonomous manipulation with a general-purpose simple hand. Intl. Journal of Robotics Research (IJRR) 31(5) (2012)

Matuk Herrera, R.: Multilayer perceptrons for bio-inspired friction estimation. In: Rutkowski, L., Tadeusiewicz, R., Zadeh, L.A., Zurada, J.M. (eds.) ICAISC 2008. LNCS (LNAI), vol. 5097, pp. 828–838. Springer, Heidelberg (2008)

McLachlan, G.J., Krishnan, T.: The EM Algorithm and Extensions. Wiley Interscience (1997)

Meeussen, W., Wise, M., Glaser, S., Chitta, S., McGann, C., Patrick, M., Marder-Eppstein, E., Muja, M., Eruhimov, V., Foote, T., Hsu, J., Rusu, R., Marthi, B., Bradski, G., Konolige, K., Gerkey, B., Berger, E.: Autonomous door opening and plugging in with a personal robot. In: Proc. of the IEEE Intl. Conf. on Robotics and Automation (ICRA), Anchorage, AK, USA (2010)

Meltzoff, A.N., Moore, M.K.: Explaining facial imitation: A theoretical model. Early Development and Parenting 6(3-4), 179–192 (1997)

Monasterio, I., Lazkano, E., Rano, E., Sierra, B.: Learning to traverse doors using visual information. Mathematics and Computers in Simulation 60(3-5), 347–356 (2002)

Monzee, J., Lamarre, Y., Smith, A.M.: The effects of digital anesthesia on force control using a precision grip. Journal of Physiology 89(2), 672–683 (2003)

Motoo, K., Fukuda, T., Arai, F., Matsuno, T.: Piezoelectric vibration-type tactile sensor with wide measurement range using elasticity and viscosity change. IEEE Sensors Journal 7(7), 1044–1051 (2007)

Murillo, A.C., Kosecka, J., Guerrero, J.J., Sagues, C.: Visual door detection integrating appearance and shape cues. Robotics and Autonomous Systems (RAS) 56(6), 512–521 (2008)

Nabeshima, C., Kuniyoshi, Y., Lungarella, M.: Adaptive body schema for robotic tool-use. Advanced Robotics 10(20), 1105–1126 (2006)

Natale, L.: Linking action to perception in a humanoid robot: A developmental approach to grasping. PhD thesis, University of Genoa, Italy (2004)

Niemeyer, G., Slotine, J.-J.: A simple strategy for opening an unknown door. In: Proc. of the IEEE Intl. Conf. on Robotics and Automation (ICRA), Albuquerque, NM, USA (1997)

Nieuwenhuisen, M., Stückler, J., Behnke, S.: Improving indoor navigation of autonomous robots by an explicit representation of doors. In: Proc. of the IEEE Intl. Conf. on Robotics and Automation (ICRA), Anchorage, AK, USA (2010)

Nishihara, H.K.: PRISM: a practical real-time imaging stereo matcher. Technical Report AIM-780, Massachusetts Institute of Technology, Cambridge, MA, USA (1984)

Nüchter, A., Hertzberg, J.: Towards semantic maps for mobile robots. Robotics and Autonomous Systems (RAS) 56(11), 915–926 (2008)

Ohmura, Y., Kuniyoshi, Y., Nagakubo, A.: Conformable and scalable tactile sensor skin for curved surfaces. In: Proc. of the IEEE Intl. Conf. on Robotics and Automation (ICRA), Orlando, FL, USA (2006)

Okamura, A.M., Cutkosky, M.R.: Haptic exploration of fine surface features. In: Proc. of the IEEE Intl. Conf. on Robotics and Automation (ICRA), Detroit, MI, USA (1999)

Omata, S., Murayama, Y., Constantinou, C.E.: Real time robotic tactile sensor system for the determination of the physical properties of biomaterials. Sensors and Actuators 112(2-3), 278–285 (2004)

O'Regan, J.K., Noë, A.: A sensorimotor account of vision and visual consciousness. The Behavioral and Brain Sciences 24(5), 939–973 (2001)

Pardowitz, M., Dillmann, R.: Towards life-long learning in household robots: The Piagetian approach. In: Proc. of the IEEE Intl. Conf. on Development and Learning (ICDL), London, UK (2007)

Parlitz, C., Hägele, M., Kleint, P., Seifertt, J., Dautenhahn, K.: Care-O-Bot 3 – Rationale for human-robot interaction design. In: Proc. of the Intl. Symp. on Robotics (ISR), Seoul, Korea (2008)

Pastor, P., Hoffmann, H., Asfour, T., Schaal, S.: Learning and generalization of motor skills by learning from demonstration. In: Proc. of the IEEE Intl. Conf. on Robotics and Automation (ICRA), Kobe, Japan (2009)

Pearl, J.: Probabilistic Reasoning in Intelligent Systems: Networks of Plausible Inference. Morgan Kaufmann (1988)

Peters, J., Vijayakumar, S., Schaal, S.: Reinforcement learning for humanoid robotics. In: Proc. of the IEEE-RAS Intl. Conf. on Humanoid Robots (Humanoids), Karlsruhe, Germany (2003)

Petrovskaya, A., Ng, A.: Probabilistic mobile manipulation in dynamic environments, with application to opening doors. In: Proc. of the Intl. Conf. on Artificial Intelligence (IJCAI), Hyderabad, India (2007)

Petrovskaya, A., Khatib, O., Thrun, S., Ng, A.Y.: Bayesian estimation for autonomous object manipulation based on tactile sensors. In: Proc. of the IEEE Intl. Conf. on Robotics and Automation (ICRA), Orlando, FL, USA (2006)

Petrovskaya, A., Thrun, S., Koller, D., Khatib, O.: Guaranteed inference for global state estimation in human environments. In: Proc. of the Workshop on Mobile Manipulation at Robotics: Science and Systems (RSS), Zaragoza, Spain (2010)

Pezzementi, Z.A., Reyda, C., Hager, G.D.: Object mapping, recognition, and localization from tactile geometry. In: Proc. of the IEEE Intl. Conf. on Robotics and Automation, ICRA (2011)

Pradeep, V., Konolige, K., Berger, E.: Calibrating a multi-arm multi-sensor robot: A bundle adjustment approach. In: Intl. Symp. on Experimental Robotics (ISER), New Delhi, India (2010)

Quigley, M., Batra, S., Gould, S., Klingbeil, E., Le, Q., Wellman, A., Ng, A.Y.: High-accuracy 3D sensing for mobile manipulation: Improving object detection and door opening. In: Proc. of the IEEE Intl. Conf. on Robotics and Automation (ICRA), Kobe, Japan (2009)

Quinlan, J.R.: Learning with continuous classes. In: Australian Joint Conf. on Artificial Intelligence, Singapore (1992)

Quinlan, J.R.: C4.5: Programs for Machine Learning. Morgan Kaufmann (1993)

Rasmussen, C.E., Williams, C.K.I.: Gaussian Processes for Machine Learning. In: Adaptive Computation and Machine Learning. The MIT Press (2006)

Ratliff, N., Bagnell, J.A., Srinivasa, S.: Imitation learning for locomotion and manipulation. In: Proc. of the IEEE-RAS Intl. Conf. on Humanoid Robots, Humanoids (2007)

Ratliff, N., Zucker, M., Bagnell, J.A., Srinivasa, S.: CHOMP: Gradient optimization techniques for efficient motion planning. In: Proc. of the IEEE Intl. Conf. on Robotics and Automation, ICRA (2009)

Reinhart, R.F., Steil, J.J.: Recurrent neural associative learning of forward and inverse kinematics for movement generation of the redundant PA-10 robot. In: Proc. of the ECSIS Symp. on Learning and Adaptive Behaviors for Robotic Systems (LAB-RS), Edinburgh, United Kingdom (2008)

Remolina, E., Kuipers, B.: Towards a general theory of topological maps. Artificial Intelligence 152(1), 47–104 (2004)

Rolf, M., Steil, J.J., Gienger, M.: Efficient exploration and learning of whole body kinematics. In: Proc. of the IEEE Intl. Conf. on Development and Learning (ICDL), Shanghai, China (2009)

Ross, D.A., Tarlow, D., Zemel, R.S.: Learning articulated structure and motion. Intl. Journal of Computer Vision (IJCV) 88(2), 214–237 (2010)

Roweis, S.T., Saul, L.K.: Nonlinear dimensionality reduction by locally linear embedding. Science 290(5500), 2323–2326 (2000)

Roy, N., Thrun, S.: Online self-calibration for mobile robots. In: Proc. of the IEEE Intl. Conf. on Robotics and Automation (ICRA), Detroit, MI, USA (1999)

Russell, R.A.: Object recognition by a 'smart' tactile sensor. In: Proc. of the Australian Conf. on Robotics and Automation, Melbourne, Australia (2000)

Rusu, R.B., Meeussen, W., Chitta, S., Beetz, M.: Laser-based perception for door and handle identification. In: Proc. of the Intl. Conf. on Advanced Robotics (ICAR), Munich, Germany (2009)

Saal, H.P., Ting, J., Vijayakumar, S.: Active estimation of object dynamics parameters with tactile sensors. In: Proc. of the IEEE/RSJ Intl. Conf. on Intelligent Robots and Systems (IROS), Taipei, Taiwan (2010)

Sawa, F., Ogino, M., Asada, M.: Body image constructed from motor and tactile images with visual information. Intl. Journal of Humanoid Robotics 4(2), 347–364 (2007)

Schaal, S., Peters, J., Nakanishi, J., Ijspeert, A.J.: Learning movement primitives. In: Proc. of the Intl. Symp. of Robotics Research (ISSR), Siena, Italy (2003)

Schneider, A.: Objektklassifikation mittels Tastsensoren. BSc thesis, University of Freiburg, Germany (2009)

Schwarz, G.: Estimating the dimension of a model. The Annals of Statistics 6(2), 461–464 (1978)

Sciavicco, L., Siciliano, B.: Modeling and Control of Robot Manipulators. Advanced Textbooks in Control and Signal Processing. Springer (2000)

Sentis, L., Park, J., Khatib, O.: Compliant control of multi-contact and center of mass behaviors in humanoid robots. IEEE Transactions on Robotics (T-RO) 26(3), 483–501 (2010)

Sinapov, J., Stoytchev, A.: The boosting effect of exploratory behaviors. In: Proc. of the National Conf. on Artificial Intelligence (AAAI), Atlanta, GA, USA (2010)

Sinapov, J., Wiemer, M., Stoytchev, A.: Interactive learning of the acoustic properties of household objects. In: Proc. of the IEEE Intl. Conf. on Robotics and Automation (ICRA), Kobe, Japan (2009)

Srinivasa, S., Ferguson, D., Helfrich, C., Berenson, D., Romea, A.C., Diankov, R., Gallagher, G., Hollinger, G., Kuffner, J., Vandeweghe, J.M.: HERB: a home exploring robotic butler. Autonomous Robots 28(1), 5–20 (2010)

Stamenov, M.I.: Body schema, body image, and mirror neurons. In: De Preester, H., Knockaert, V. (eds.) Body Image and Body Schema: Interdisciplinary Perspectives on the Body, ch. 2. John Benjamins Publishing (2005)

Takamuku, S., Fukuda, A., Hosoda, K.: Repetitive grasping with anthropomorphic skin-covered hand enables robust haptic recognition. In: Proc. of the IEEE/RSJ Intl. Conf. on Intelligent Robots and Systems (IROS), Nice, France (2008)

Taycher, L., Fisher, J.W., Darrell, T.: Recovering articulated model topology from observed rigid motion. In: Proc. of the Conf. on Neural Information Processing Systems (NIPS), Vancouver, Canada (2002)

Tenenbaum, J.B., de Silva, V., Langford, J.C.: A global geometric framework for nonlinear dimensionality reduction. Science 290(5500), 2319–2323 (2000)

Thrun, S., Burgard, W., Fox, D.: Probabilistic Robotics. The MIT Press (2005)

Ting, J., Mistry, M., Peters, J., Schaal, S., Nakanishi, J.: A Bayesian approach to nonlinear parameter identification for rigid body dynamics. In: Proc. of Robotics: Science and Systems (RSS), Philadelphia, PA, USA (2006)

Torr, P.H.S., Zisserman, A.: MLESAC: A new robust estimator with application to estimating image geometry. Computer Vision and Image Understanding 78(1), 138–156 (2000)

Tso, S.K., Liu, K.P.: Hidden Markov model for intelligent extraction of robot trajectory command from demonstrated trajectories. In: Proc. of the IEEE Intl. Conf. on Industrial Technology (ICIT), Shanghai, China (1996)

Ueda, J., Ishida, Y., Kondo, M., Ogasawara, T.: Development of the NAIST-hand with vision-based tactile fingertip sensor. In: Proc. of the IEEE Intl. Conf. on Robotics and Automation (ICRA), Barcelona, Spain (2005)

Ware, R., Lad, F.: Approximating the distribution for sums of products of normal variables. Technical Report UCDMS 2003/15, University of Canterbury, New Zealand (2003)

Wedel, A., Rabe, C., Vaudrey, T., Brox, T., Franke, U., Cremers, D.: Efficient dense scene flow from sparse or dense stereo data. In: Forsyth, D., Torr, P., Zisserman, A. (eds.) ECCV 2008, Part I. LNCS, vol. 5302, pp. 739–751. Springer, Heidelberg (2008)

Weiss, K., Wörn, H.: The working principle of resistive tactile sensor cells. In: Proc. of the IEEE Intl. Conf. on Robotics and Automation (ICRA), Barcelona, Spain (2005)

Wieland, S., Gonzalez-Aguirre, D., Vahrenkamp, N., Asfour, T., Dillmann, R.: Combining force and visual feedback for physical interaction tasks in humanoid robots. In: Proc. of the IEEE-RAS Intl. Conf. on Humanoid Robots (Humanoids), Paris, France (2009)

Williams, C., Shang, D., Carnahan, H.: Pressure is a viable controlled output of motor programming for object manipulation tasks. In: Kappers, A.M.L., van Erp, J.B.F., Bergmann Tiest, W.M., van der Helm, F.C.T. (eds.) EuroHaptics 2010, Part II. LNCS, vol. 6192, pp. 339–344. Springer, Heidelberg (2010)

Xu, Z., Deyle, T., Kemp, C.C.: 1000 trials: an empirically validated end effector that robustly grasps objects from the floor. In: Proc. of the IEEE Intl. Conf. on Robotics and Automation, ICRA (2009)

Yan, J., Pollefeys, M.: Automatic kinematic chain building from feature trajectories of articulated objects. In: Proc. of the IEEE Conf. on Computer Vision and Pattern Recognition (CVPR), New York, NY, USA (2006)

Yoshikawa, Y., Hosoda, K., Asada, M.: Binding tactile and visual sensations via unique association by cross-anchoring between double-touching and self-occlusion. In: Proc. of the Intl. Workshop on Epigenetic Robotics, Genoa, Italy (2004a)

Yoshikawa, Y., Tsuji, Y., Hosoda, K., Asada, M.: Is it my body? Body extraction from uninterpreted sensory data based on the invariance of multiple sensory attributes. In: Proc. of the IEEE/RSJ Intl. Conf. on Intelligent Robots and Systems (IROS), Sendai, Japan (2004b)

Yufeng, Y.L., Emery, R., Chakrabarti, D., Burgard, W.: Using EM to learn 3D models of indoor environments with mobile robots. In: Proc. of the Intl. Conf. on Machine Learning (ICML), Williamstown, MA, USA (2001)

Zhang, J., Marszalek, M., Lazebnik, S., Schmid, C.: Local features and kernels for classification of texture and object categories: A comprehensive study. Intl. Journal of Computer Vision (IJCV) 73(11), 123–138 (2007)

Ziebart, B.D., Maas, A., Bagnell, J.A., Dey, A.K.: Maximum entropy inverse reinforcement learning. In: Proc. of the National Conf. on Artificial Intelligence (AAAI), Chicago, IL, USA (2008)

Index

MIX
Papier aus verantwortungsvollen Quellen
Paper from responsible sources
FSC® C105338

If you have any concerns about our products,
you can contact us on
ProductSafety@springernature.com

In case Publisher is established outside the EU,
the EU authorized representative is:
Springer Nature Customer Service Center GmbH
Europaplatz 3, 69115 Heidelberg, Germany

Printed by Libri Plureos GmbH
in Hamburg, Germany